《不一样的衣服》编委会

主　编：赵石定

副主编：周　祥　段兴民

执行副主编：马　滨　李　爽

编　委：王　梅　张益珲　孙　琦

　　　　范龙坤　谢牧耘　（英）威廉·威尔克斯（William Willcox）

新媒体制作：张益珲　张丽园　施俊龙

配　音：赵　纳

不 一 样 的 衣 服

CHINA: 56 INDIGENOUS COSTUMES

云南人民出版社◎编

文字：孙　琦　范龙坤

绘画：谢牧耘

翻译：（英）威廉·威尔克斯（William Willcox）

配音：赵　纳

云南出版集团

云南人民出版社

图书在版编目（CIP）数据

不一样的衣服：汉、英 / 云南人民出版社编；孙琦, 范龙坤文字撰稿；谢牧耘绘画；(英) 威廉·威尔克斯 (William Willcox) 翻译. —昆明：云南人民出版社, 2021.1

ISBN 978-7-222-19736-7

Ⅰ.①不… Ⅱ.①云… ②孙… ③范… ④谢… ⑤威… Ⅲ.①少数民族—民族服饰—云南—图集 Ⅳ. ①TS941.742.8-64

中国版本图书馆CIP数据核字(2020)第196050号

出 版 人：赵石定
责任编辑：李 爽 张益珲
装帧设计：马 滨 王冰洁
责任校对：任建红 张丽园 刘振芳
数字编辑：张益珲 张丽园 施俊龙
责任印制：代隆参

不一样的衣服
CHINA：56 INDIGENOUS COSTUMES

云南人民出版社◎编
文字：孙 琦 范龙坤
绘画：谢牧耘
翻译：（英）威廉·威尔克斯（William Willcox）
配音：赵 纳

出版 云南出版集团 云南人民出版社
发行 云南人民出版社
社址 昆明市环城西路609号
邮编 650034
网址 www.ynpph.com.cn
E-mail ynrms@sina.com
开本 889mm×1194mm 1/16
印张 7.5
字数 60千
版次 2021年1月第1版第1次印刷
印刷 重庆新金雅迪艺术印刷有限公司
书号 ISBN 978-7-222-19736-7
定价 118.00元

如需购买图书、反馈意见，请与我社联系
总编室 0871-64109126 发行部 0871-64108507
审校部 0871-64164626 印制部 0871-64191534

云南人民出版社微信公众号

CHINA: 56
INDIGENOUS
COSTUMES

目录

汉族 / Han / 002

回族 / Hui / 004

柯尔克孜族 / Kirgiz / 006

塔吉克族 / Tajik / 008

乌孜别克族 / Ozbek / 010

俄罗斯族 / Russian / 012

哈萨克族 / Kazak / 014

维吾尔族 / Uyghur / 016

塔塔尔族 / Tatar / 018

撒拉族 / Salar / 020

土族 / Tu / 022

裕固族 / Yugur / 024

东乡族 / Dongxiang / 026

保安族 / Bonan / 028

藏族 / Tibetan / 030

珞巴族 / Lhoba / 032

门巴族 / Moinba / 034

羌族 / Qiang / 036

彝族 / Yi / 038

傣族 / Dai / 040

阿昌族 / Achang / 042

景颇族 / Jingpo / 044

德昂族 / De'ang / 046

怒族 / Nu / 048

独龙族 / Drung / 050

傈僳族 / Lisu / 052

纳西族 / Naxi / 054

普米族 / Pumi / 056

白族 / Bai / 058

佤族 / Va / 060

布朗族 / Blang / 062

哈尼族 / Hani / 064

拉祜族 / Lahu / 066

基诺族 / Jino / 068

蒙古族 / Mongol / 070

土家族 / Tujia / 072

苗族 / Miao / 074

侗族 / Dong / 076

布依族 / Bouyei / 078

仡佬族 / Gelo / 080

仫佬族 / Mulam / 082

水族 / Shui / 084

满族 / Manchu / 086

达斡尔族 / Daur / 088

鄂温克族 / Ewenki / 090

鄂伦春族 / Oroqen / 092

赫哲族 / Hezhe / 094

朝鲜族 / Korean / 096

锡伯族 / Xibe / 098

壮族 / The Zhuang / 100

瑶族 / Yao / 102

毛南族 / Maonan / 104

京族 / Jing / 106

黎族 / Li / 108

畬族 / She / 110

高山族 / The Gaoshan / 112

编者注：

　　本书民族排序方式为自西向东，自北向南，将整个中国分为五个板块：西北、西南、中部、东北、东南。原则上先出现的民族先呈现，但如遇到某一民族主体所在地顺序靠后，就放在后面呈现。

EDITOR'S NOTE

　　The order of ethnic groups in this book is based on China's geography, from west to east and north to south, and it includes 5 regions: northwest, southwest, middle, northeast and southeast. Therefore, ethnic groups that reside in the first of the aforementioned regions will appear first and in sequence according to the area they mostly inhabit.

要问古时候的事，老辈子已绣在衣服上了；要分辨各个民族兄弟姐妹，看看穿着的和挂着的就清楚啦！

——彝族民歌

Regarding ancient stories, these were embroidered by ancestors on our clothes; To distinguish between ethnic groups, look at what they wear.

— Yi folksong

汉族
Han

中国人口最多的民族，全国各地均有分布。汉族服饰具有博采兼容的气度：西北的羊肚巾、红肚兜和敞怀对襟衣潇洒风流；西南的大盘头、花围腰和绣花衣折射了多民族文化和谐共处的光彩；东北的老棉袄、旗袍和马褂透出历史的底蕴；沿海客家人的衫裤与凉笠吹拂着大海的气息……不同礼俗、不同行业有不同的传统形式；改革开放更使汉装兼收并蓄，盛放一片花潮。

The Han people are the most populous ethnic group in China and they are spread across all regions. The Han costume is all embracing: The northwestern-style turbans, red undergarments and unbuttoned jackets are unencumbered and refined. The pan-head hairstyle, flower aprons and embroidered clothing of Southwest China reveal the radiant harmony of multi-ethnicity. The cotton-quilted jackets, qipao and magua(A short garment worn over a robe) of Northeast China reveal the details of history. The clothes and straw hats of the coastal Hakka people are kissed by the sea. Different customs and occupations led to the diversification of traditional garments , after the reform and opening-up policy of China, these aesthetics have been integrated in Han clothing, where they flourish.

时代风华，尽显绅士气度
A Gentleman's Fashion

民国时期的汉族男子服装，以长衫、马褂为代表。初期基本上保持着清代旧制，后受西方国家的影响，也开始穿着西装，但并不排斥原来的服饰，长

衫、马褂仍被用作礼服，与西装革履并行不悖。

The iconic costume worn by male Han people during the Minguo period (1912-1949) consisted of a Mandarin jacket and long gown. These garments continued the style during the Qing dynasty .However, with the influence of Western fashion, the popularity of suits started to flourish. Western suits however did not replace traditional clothing outright, because Mandarin jackets and long gowns are still used as ceremonial dress and do not conflict with western clothes.

优雅却不失东方神韵的高贵
The Elegant and Charming Qipao

民国最有影响力的女装当为旗袍。清末民初，汉族妇女受满族影响穿着旗袍并形成时髦，但民国的旗袍在传统基础上加以改进，最终演变成为一种展现东方女性气质、具有独特风貌的服装，并享有"国服"之誉。

The qipao was the most popular choice of garment during the Minguo period. The traditional qipao, which had became fashionable with the Manchu influence during the late Qing dynasty and early Minguo period, was improved upon by transforming it into a costume that can embody both oriental femininity and uniqueness of style. It now enjoys the reputation of being China's national costume.

回族
Hui

回族是中国人口较多的一个少数民族，全国各地均有分布。宁夏回族自治区是其主要聚居区，北京、河北、内蒙古、辽宁、安徽、山东、河南、云南、甘肃、青海和新疆等地回族也较多。回族通常称服饰为"衣着""穿戴"。居住在甘肃、宁夏、青海地区的回族服饰保留了较多本民族传统，其他地区则普遍入乡随俗。如在汉族地区，回民服饰与汉族服装大体相同；在新疆的部分回族穿维吾尔族服饰；在西藏的部分回族穿藏族服饰；在云南的部分回族穿彝、傣等民族服饰；在贵州的部分回族则穿苗族服饰等。

The Hui are a relatively populous ethnic group that reside in all regions of China. A large portion live in Ningxia Hui Autonomous Region. There are also many Hui in Beijing, Hebei, Inner Mongolia, Liaoning, Anhui, Shandong, Henan, Yunnan, Gansu, Qinghai and Xinjiang. Among Hui people, clothing is frequently referred to as "dressing" or "wearing". The dress worn by the Hui people living in Gansu, Ningxia and Qinghai has retained much of their original styles, while those in other regions have adapted to follow local norms. For example, the dress of the Hui people living in the Han regions is for the most part the same as the Han clothing; Some Hui people living in Xinjiang wear Uyghur clothes; Some Hui people living in Tibet wear Tibetan clothes; Some Hui people in Yunnan wear Yi and Dai ethnic clothes and Some Hui people living in Guizhou wear Miao clothes.

长在头上的方块冰淇淋
Small White and Black Hats

最具回族特色的小白帽，叫"号帽"，又称"顶帽"。多用白布制作，或用白线钩成；黑帽则多用华达呢、呢绒制作，或用黑毛线编结，简洁又大气。

The small white hat which characterises the Hui is known as a horn hat, a top hat and it is made of white cloth or crocheted from white thread. There are also black hats and these are mostly made of gaberdine, woollen fabric or knitted using black wool. It looks elegant and modest.

多功能蒙面神器"古古"
Headscarves: *Gugu*

回族女性的遮面护发头巾，又称作"古古"。用丝、绸、纱、绒等精细面料制成，呈筒形，用时从头上套下，能把头发、半个面部、脖颈全部遮住，保暖、防晒又神秘美丽。

The headscarves Hui women use to cover their hair are known by the name *Gugu*. Made of fine fabrics such as silk, satin, yarn or velvet, the gugu forms a loop of fabric that can cover the head, lower face and neck. This is a warm, sun-proof and beautiful garment.

扫码收听 音频内容

柯尔克孜族
Kirgiz

　　跨境游牧民族，也是吉尔吉斯斯坦的主体民族，中国境内主要聚居在新疆维吾尔自治区孜勒苏柯尔克孜自治州，新疆其他地区和黑龙江省也有散居。大部分柯尔克孜族逐水草而居，其服饰多以动物毛皮及家织毛布制作。男子以袍服为主，女子以裙装居多。生性乐观的柯尔克孜人喜欢演奏乐器，爱过节，节日里，荡秋千的姑娘长裙飞扬，头巾鲜艳，赛马刁羊的小伙子则穿对襟无纽扣绣花短衣，绣花长裤扎在长筒革靴中，头戴顶白檐黑的翻檐毡帽，风度翩翩。

　　The Kirgiz is the main ethnic group in Kyrgyzstan and those living in China are mostly concentrated in Zilesu Kirgiz Autonomous Prefecture of Xinjiang with some scattered across other regions of Xinjiang and Heilongjiang. Most Kirgiz people live where there is water and pastureland and their clothing is generally made of fur and homespun wool, with men mainly wearing robes and women skirts. The optimistic Kirgiz people enjoy playing musical instruments and festivities. During a festival, the long skirts of girls swinging their hips can be seen flying alongside bright headscarves. Smart and elegant competing men wear embroidered buttonless shirts, long embroidered trousers that are tucked into long leather boots, and black felt hats which have white upturned brims.

硬派"花样"男子
Tough Men in Patterned Clothing

　　柯尔克孜族男子外衣上束皮带或

绣花腰带,足蹬高筒马靴或皮鞋,整个装束在显示了草原民族粗犷气概的同时,也不乏精悍利落,有自己的特色。

In terms of outerwear, Kirgiz men wear a leather or embroidered belt and riding boots or leather shoes. This outfit shows the bold and unconstrained spirit of a grassland group with its own characteristics.

华丽丽的"叶列切克"
The Magnificent *Elechek Hat*

传统的柯尔克孜族女帽由软帽和缠于其上的白色丝绸组成,是已婚妇女的标志。两帽耳挂有珊瑚珠和银片、银盒等坠饰,垂至膝盖;帽子正面饰有红宝石和珊瑚珠串流苏,极其奢华。

The *Elechek* is a traditional hat worn by married Kirgiz women and is a soft cap wrapped around by white silk. Hanging from the two ears of the hat to knees are coral beads, silver pendants and silver boxes. At the front of this hat are opulent tassels decorated with rubies and coral beads.

扫码收听 音频内容

塔吉克族
Tajik

　　跨境民族，主要分布在中亚的塔吉克斯坦等国，中国境内的塔吉克族多数聚居于新疆塔什库尔干塔吉克自治县。生活在高海拔地区的塔吉克族，喜穿坎肩、长裙、马靴。帽子是他们较为独特的服饰。男女各有不同，不仅美观大方，还能防风保暖。塔吉克人喜欢鲜艳的红色，在冰雪皑皑的高山上十分耀眼，也给人一种温暖感。订婚时男方所送的礼物中必有大红方巾；新娘穿红色长裙和红皮靴，外罩红色袷袢；新郎的帽子外面也缠以红白两色相间的彩布，红色表示青春，白色表示纯洁。

　　The Tajik people mostly live in Central Asian countries including Tajikistan. Those in China are mostly concentrated in Tashkurgan Tajik autonomous county of Xinjiang. Living at high-altitude, they like to wear sleeveless jackets, long skirts and riding boots. The hat is one of the more unique aspects of their costume and those worn by men differ from women, not only in terms of beauty, also with regards to whether they are windproof and warm. The Tajik people like bright red, a warming colour which is noticeable in the snow-capped mountains. During a couple's engagement, the man must give a large red scarf as a gift and the bride wears a long red dress, red leather shoes and a red chapan. The groom's hat is also wrapped with red and white coloured cloth. Red means youth and white means purity.

塔吉克男帽——北疆生活必备
The Male Tajik Hat — A Must Have for Life in Northern Xinjiang

　　塔吉克族的男帽呈圆高筒形，

多以黑平绒布为面，黑羊羔皮为里。此皮帽戴时可自由翻卷，平时将帽檐卷起，露出一圈皮毛里子作为装饰，遇风沙时则将帽檐拉下，遮住面颊和双耳。

The hats worn by Tajik men are rounded and cylindrical and their exteriors are made of black velveteen while the interiors are made of black lambskin. The leather hat can be rolled freely when it is worn and normally the brim is rolled up to reveal a circle of fur lining for decorative purposes. In case of wind or sand, the brim can be pulled down to cover the cheeks and ears.

圆筒女帽——激情燃烧的岁月
The Female Tajik Hat

女帽亦为圆筒形，但华美程度甚于男帽。少女的圆帽用紫、金黄、大红色的平绒精心绣制，帽前檐垂饰色彩艳丽的串珠或小银链，外出时，帽上再加披大方巾，还能护住双肩及前额。

While the hats worn by Tajik women are also cylindrical, they are more resplendent. The hats worn by girls are meticulously embroidered with purple, golden or red velveteen and colourful beads or small silver chains decorate the front brim. A large square cloth, which can protect the forehead and shoulders, is used to cover the hat when outside.

乌孜别克族
Ozbek

跨境民族，主体在中亚的乌兹别克斯坦等国，中国境内的乌孜别克族散居在新疆的喀什、莎车、叶城、乌鲁木齐、伊宁、木垒等地。乌孜别克人一般均穿皮靴，外加浅帮套鞋。绣花女皮靴工艺精细，是著名手工艺品。乌孜别克族无论男女老幼，都戴绣花小帽或灯芯绒素面小帽。年轻妇女喜穿宽大多褶的连衣裙，脚穿绣花鞋，头戴小花帽；老年妇女服装则偏深色（墨绿、咖啡或黑），头扎白巾。男子服装，为绣花小立领白衫，长裤、筒靴，外罩竖条敞胸长衫。

While mainly living in Uzbekistan, some Ozbek people live in Xinjiang province of China. Ozbek men usually wear leather boots under short overshoes. The finely crafted women's embroidered leather boots are a famous handicraft. Everyone wears an embroidered or corduroy cap, irrespective of whether they are male or female, young or old. Young women like to wear wide pleated dresses, embroidered shoes and colourful hats while older women tend to wear darker colours (dark green, brown or black) and white headscarves. Men wear embroidered white straight-collared shirts, trousers, boots and a long open-breasted robe.

"托尼"长袍，低调奢华的冬日之选
Men's Robe — Winter Choice

乌孜别克族男子常服称为"托尼"。过去多用质地较厚的绸缎、金

丝绒或青、黑色厚布制作，现在多选用各种毛料。长袍边缘绣有红、黄、绿色纹样。青年男子的长袍颜色较鲜艳，老年人的长袍以深色为主。

0zbek men often wear clothes called *Tony*. In the past, these were made of thick satin, gold velvet or thick green and black cloth, but nowadays a variety of woollen materials are generally used. The robes are embroidered with red, yellow and green patterns and those worn by young men are brightly coloured while darker colours are used in those worn by older men.

"魅依纳克"，女孩们的四季暖阳
Ozbek Dresses

乌孜别克族女性喜欢穿连衣裙，当地称其为"魅依纳克"。年轻女孩的连衣裙宽大多褶，花色亮丽，裙褶呈波浪状，摆动起来别有情趣。不束腰带时在裙外加穿绣花小坎肩。

Ozbek women like to wear dresses. Young girls usually wear bright and pretty roomy pleated dresses , because the pleats are wavy, they are particularly interesting when they sway. An embroidered waistcoat is worn when a belt is not.

扫码收听
音频内容

俄罗斯族
Russian

　　跨境民族，中国境内主要聚居在新疆伊犁地区，部分散居在新疆的塔城、阿勒泰、乌鲁木齐以及黑龙江省和内蒙古自治区等地。中国的俄罗斯族历经百年岁月，其外貌、长相、风俗和习惯等，已渐渐形成了自己的特色。俄罗斯族的服饰丰富多彩，会在不同季节里选择不同款式的衣着。年轻人爱穿各种时装。男子夏季多穿丝绸白色直领汗衫长裤，春秋穿茶色或铁灰色西装，冬季穿翻领皮大衣或棉衣，脚穿高筒皮靴或毡靴。妇女夏季多穿短袖、印花的连衣裙，春秋穿西服上衣或西服裙，头戴彩色呢礼帽，冬季穿裙子，外套半长皮大衣，脚穿高筒皮靴，头戴毛织大头巾。

　　The Russian people living in China are mostly settled in the Yili Region of Xinjiang with some scattered in Tacheng, Altay and Urumqi of Xinjiang, Heilongjiang and Inner Mongolia. After years of assimilation, the Russian people in China have gradually formed their own characteristics in appearance, features, customs and habits. Russian costumes are both rich and colourful, and different styles of clothing are selected in different seasons. Young people love to wear all kinds of fashion. In summer, men often wear white silk straight-collared vests and trousers, a brown or grey suit in spring and a leather or cotton-padded turndown collar coat in winter with tall leather or felt boots. Women usually wear short-sleeve printed dresses in summer, a suit jacket or shirt and a coloured woollen formal hat in spring and autumn and, in winter, a blouse, a half-length leather overcoat, tall leather boots and a large woollen headscarf.

优雅王子的精致衬衫
The Elegant and Exquisite Shirt

传统的俄罗斯族男子夏季多穿白色衬衫，特点是整体宽松，两袖筒肥大，半开襟立领，穿时由头上套下。衣上镶有花边，上绣精美细密的几何或花草图案，十足王子范儿。

Traditional Russian men mostly wear white shirts in summer. These shirts are characteristically loose-fitting all-over, including the roomy sleeves, and have half-opened lappets and upright collars. The shirt is donned by pulling it over the head. The shirt is decorated with lace and embroidered with fine geometric or flower patterns, making it look princely.

四季都是爆款的"布拉吉"
Dresses are a Success All Year Round

传统俄罗斯族女性冬夏均穿名为"布拉吉"的连衣裙，冬季常在连衣裙外加罩一件无袖长袍。年轻女孩的无袖长袍多在两侧开衩。如内穿浅色连衣裙，则配穿深色袍。

Traditional Russian women wear dresses in winter and summer. A long sleeveless robe is often worn above the dress in winter and those worn by young girls are often slit on both sides. If a light-coloured dress is worn, then the norm is to choose a dark robe.

扫码收听音频内容

哈萨克族
Kazak

　　跨境民族，主体在中亚、西亚，中国境内主要聚居在新疆的伊犁、木垒、巴里坤，甘肃阿克塞等地。哈萨克族服饰一般多为手工制作，以羊皮毛、山羊绒、狼皮、狐狸皮等为材料，他们还赋予了服饰色彩一定的象征意义。如：绿色象征草原，白色象征乳汁与羊群，红色象征阳光，黄色象征人类生存的大地。哈萨克女装以华美尊贵著称，帽羽多用猫头鹰毛，帽边有华丽镶绣，紧身连衣裙下摆及袖口缝制三层飞边皱褶，娇似公主。当你在"姑娘追"等民俗活动中，看到头插轻柔羽毛、跃马扬鞭的姑娘，那一定是哈萨克族。男子服装多为新疆地区流行式样，追求潇洒大方，显示草原男子的粗犷、豪放气概。式样各异的帽子是哈萨克服饰的一个亮点。

　　The Kazak people living in China are mostly situated in Xinjiang and Gansu. The Kazak costumes are generally handmade with sheepskin, cashmere, wolf skin and fox skin as materials and a certain symbolic meaning is given by the colour. For example, green symbolises the grasslands, white the milk and the sheep, red symbolises sunlight, yellow symbolises the lands of human existence. Clothes worn by Kazak women are famed for their resplendence. Owl feathers are mostly used to decorate hats that have magnificent embroidered brims. The hems and cuffs of the tight-fitting dresses are sewn with three pleated layers, which are as delicate as a princess' dress. When you see a girl wearing soft feathers in her hair while galloping on horseback, you will know she is Kazak. The clothing worn by Kazak

men in Xinjiang is most fashionable, which shows bold spirit of the grassland men. A highlight of the Kazak costume comes in the form of their various hats that they wear.

高颜值的卷边毡帽
Rolled-up Felt Hat

伊犁地区的哈萨克族男性，夏季多戴用细白毡制成的毡帽，帽檐上卷，四周镶黑边，帽顶呈方形，洁白耀眼，引人注目，不仅拉高颜值，还是增高神器。

In the summer months, Kazak men of Yili area wear white hats made of fine felt that has a rolled up brim and black edges. This spotless dazzling hat is most eye-catching and, not only does it enhance the wearer's good looks, it also makes him look taller.

新娘专属的"沙吾克烈"凤冠帽
The Bride's Phoenix Crown

这种帽子是姑娘出嫁时所戴，尖顶，帽上绣花，并用金、银、珠宝装饰，帽子正前方还饰有串珠。作为新娘的标志，婚后要戴一年，一年后改戴头巾，生下第一个孩子后开始戴披巾。

The phoenix crown is a hat worn by a girl when she marries. It is pointed, embroidered, decorated with gold, silver and jewels and has a string of beads adorning the front. This hat is worn for one year after the wedding day as a symbol of marriage. Afterwards, it is replaced with a headscarf and a shawl is worn after the woman gives birth to her firstborn.

维吾尔族
Uyghur

　　跨境民族，中国境内主要分布在新疆维吾尔自治区。天山以南塔里木盆地周围的绿洲，以及天山东端的吐鲁番盆地是维吾尔族的聚居中心。此外，湖南省桃源县、河南省渑池县也有少量维吾尔族分布。维吾尔族服饰形式清晰，纹饰多样，色彩鲜明，图案古朴，工艺精湛。男子特色服装为对襟条纹长衫，腰间扎方形围巾，内衣侧开领，穿皮靴，戴有棱的小花帽"朵巴"；女子多着翻领分段丝绸长衫或连衣裙，色泽鲜艳，衣料轻柔，上身套一深色紧身坎肩，张弛有度，十分潇洒秀丽。

　　The Uyghur living in China are mostly settled in Xinjiang. Uyghur costumes are distinctive in form, consist of diverse patterns, bright colours, classical designs and are exquisitely made. The clothing that characterises Uyghur men consists of a buttoned striped gown with a square scarf tied around the waist, underclothing with a side-opening collar, leather boots and a small patterned cap known as a *doppa*. Women often wear a brightly-coloured and soft silk two-piece gown or dress under a tight-fitting dark-toned sleeveless jacket.

"裕袢"——最便捷的长袍
Chapan — The Most Convenient Robe

　　最具代表性的维吾尔族服装式样。男女均穿，男性穿着为多，袍长及膝或过膝，对襟，用长方巾束扎，这种随意性很强的长方巾既可束身又舒畅保暖，给惯于骑马的维吾尔人带来了

极大的方便。

Chapan is the most representative Uyghur clothing and, while it is worn by either man and woman, it is more common among males. The robe descends to around or below knee height and is buttoned down the front. A long scarf is also tied for warmth, bringing great convenience to the horseback-riding Uyghurs.

艾德莱丝绸——维吾尔服装的灵魂

Etles Silk—The Soul of Uyghur Clothes

艾德莱丝绸是维吾尔族女性服装的主要材料。制作时先采用扎染法按图案要求在经线上扎结进行染色，然后再织成面料。新疆和田、洛浦的艾德莱丝绸讲究黑白相间，空间虚实得当，布局得体；莎车的则以纹样结构细密、色彩鲜明而著称。

Etles silk is the main fabric used in the clothing worn by Uyghur women. During the production process, a method known as tie-dyeing is used to make patterned fabric. With regards to the Etles silk of Hetian and Luopu counties in Xinjiang, emphasis is put on the alternation of black and white, appropriateness of spacing and the suitability of composition, while the silk of Shache is famed for its fine and closely woven patterns and bright colours.

扫码收听
音频内容

塔塔尔族
Tatar

　　跨境民族，中国境内主要散居在新疆维吾尔自治区的伊宁、塔城、霍城、布尔津、奇台等地。塔塔尔族服饰接近欧洲传统式样。女子头裹纱巾，与头发一起向后拢扎，身穿长裙，窄袖短衣，绣花坎肩，腰系围裙。男子喜欢穿白色圆领衣，领口、袖口、胸前有十字绣图案，色彩和谐大方；下穿青布长裤，外套黑色齐腰的镶边背心。白衣、青裤，不仅色彩对比强烈，并有特定的含义：白色表示乳汁、羊群，青色代表放牧的空间——山峦、蓝天，这是对大自然的崇拜。

　　The Tatar people living in China are mainly scattered in Yining, Tacheng, Huocheng, Buerjin of Xinjiang. Tatar clothes are close to traditional European styles. Women wrap their hair in a gauze kerchief and tie it back with their hair. They also wear long dresses, narrow-sleeved tops, embroidered sleeveless jackets and aprons. Men wear white shirts with rounded collars that have cross-stitch patterns on the collars, cuffs and chests, alongside cyan trousers and a black sleeveless waistcoat. Not only do white tops and green trousers have a strong colour contrast, they also have a specific meaning. White represents milk and sheep while green represents the space for grazing, mountain ranges and the blue sky. This is worshipping mother nature.

甜美灵动的公主裙
Pretty Dresses

　　塔塔尔族女孩头戴镶珠小花帽，有的在帽子外披头巾。身穿过膝连衣裙，裙子胸部缝缀细小皱褶，裙摆宽

松、带褶，有宽大的荷叶边；袖长及腕，上部紧束，下部亦有荷叶边。连衣裙款式甜美别致，跳起舞来灵动夺目，如公主一般。

Tatar girls wear beaded flower hats and some wear headscarves outside the hat. A dress which reaches knee height is worn. The dress has small pleats stitched on the chest and the skirt hem is loose and pleated with large ruffles. The wrist-length sleeves are tight at the top and have ruffles at the bottom. This luscious dress is agile when the wearer dances.

撒拉族
Salar

主要聚居在青海省循化撒拉族自治县和化隆回族自治县甘都乡，以及甘肃省积石山保安族东乡族撒拉族自治县的大河家。撒拉族服饰与其族源和信仰有着十分密切的关系。撒拉族先民在13世纪中叶由中亚撒马尔罕迁徙到今青海省循化定居，早期本是鲜明的中亚游牧民族风格，经过与当地回族、汉族等服饰文化的融合，逐渐形成今天独特的撒拉族服饰。在色彩上男子以白、黑色为主，忌讳红、黄色及花色繁缛的服饰；妇女除参加宗教仪式外，日常衣裤鲜艳多彩，再套上黑或紫色的坎肩，更是妩媚俊俏。

The Salar people mainly live in Xunhua, Hualong of Qinghai and Dahejia of Gansu. Their clothes are closely connected to the Salar's origins and beliefs. Their ancestors migrated from Samarkand in Central Asia to Xunhua in Qinghai province in the middle of the 13th century. Their style in the early days was distinctively one of the nomadic people from Central Asia, however, after having integrated with the local Hui and Han people, the unique Salar costume seen today was gradually formed. Men mostly wear white and black and avoid red, yellow and overelaborate colours and designs. Apart from attending religious ceremonies, women's everyday clothes are colourful and they wear a charming black or purple sleeveless jacket.

送给心上人的手工爱心围肚
Making a Belt for One's Lover

撒拉族男子内衣，流行于青海循化等地。多用红、蓝、绿等彩色缎布缝制，上绣各种花卉图案。姑娘多在

婚前精心绣制围肚，结婚时将其作为礼品送给新郎，表现恩爱之心并显示刺绣手艺。

The underwear worn by Salar men is popular in Xunhua of Qinghai and other places. It is made of red, blue and green satin that is embroidered with various floral patterns. Before marriage, a girl will often embroider a belt to give as a gift to her groom when she marries, which is a way of showing both love and embroidering skills.

让脚底生花的绣花袜子
Flower Socks

撒拉语叫"吉杰合寮恩"。用黑、蓝布料密密缝成袜底后，用各色丝线在袜底绣上梅花、牡丹、葡萄等花果图案，黑底红花，色彩鲜艳，针脚细密，颇具匠心。此外还在袜底制一块"凸"字形的绣花袜跟。此种袜子尤其在婚礼的"摆针线"环节时，女方一定要向男方家人赠送。

Black and blue fabric has been used to densely stitch the bottom of the socks, and then the patterns of plum blossom, peonies and grapes are embroidered using various colours of silk thread. The red flowers on the black base, together with the bright colours and the compact stitching, show real ingenuity. In addition, a convex heel is also embroidered. This kind of sock must be given to the groom's family during the wedding.

土族
Tu

　　中国西北一个历史悠久的民族，分布于青海东部和甘肃部分地区。如果你看到有谁把彩虹的颜色放在衣袖上，那就一定是土族了。土族服饰以色彩鲜艳、对比强烈、用色大胆以及装饰精美著称。最具特色的土族妇女服饰是"花袖衫"，通常有七彩和五彩两种，土族语称"秀苏"，双袖由红、橙、蓝、白、黄、绿、黑七色或红、黄、绿、紫、蓝五色彩缎逐段镶接而成，鲜艳夺目，美观大方。男子以无袖大襟长袍和白袄为基本服装。男女都喜欢戴装饰有织锦的上翘翻檐毡帽或船形帽。

　　The Tu people have a long history in northwestern China and are distributed in eastern Qinghai and parts of Gansu. A sure sign someone is Tu is if they have the rainbow colours on their sleeves. Tu costumes are famous for their bright and bold colours, strong contrast and exquisite decorations. The *flower-sleeve blouse* is the most distinctive item of clothing worn by women. This blouse, which is known in the Tu language as *xiusu*, can be separated into two types, one consisting of seven colours and the other five. The resplendent and elegant sleeves are made by piecing together red, orange, blue, white, yellow, green and black or yellow, green, purple and blue satin. The basic clothing worn by men consists of long sleeveless robe and white jacket. Both men and women like to wear brocaded upturned felt hats and boat-shaped hats.

动静皆宜的两种帽子
Two kinds of Hat
　　土族男女都戴红缨帽和"鹰嘴啄食"白毡帽。红缨帽是一种由织锦镶

边的圆筒形毡帽。"鹰嘴啄食"白毡帽，其样式为帽子的后檐向上翻，而前檐向前展开，仿佛雄鹰正在啄食的模样。

Tu men and women wear white felt hats, which are known as *hawk-billed (food) peckers*, and red-tasseled hats. The red-tasselled hat is made of felt, cylindrical in shape and bordered by brocade. The rear brim of the *hawk-billed pecker* is turned up while the front brim expands outwards, just as if a hawk is pecking.

衣袖上随身携带的彩虹
A Rainbow Carried on the Sleeve

土族女子彩虹般的"七彩袖"极具特色，且七种色彩都有寓意。从最底层起，第一道黑色象征土地；第二道绿色象征青苗、青草；第三道黄色象征麦垛；第四道白色象征甘露；第五道蓝色象征蓝天；第六道橙色象征金色的光芒；第七道红色象征太阳。

The *seven-colour sleeves* worn by Tu women are not only distinctive, each of the colours has a meaning. Starting form the bottommost layer, black symbolises the land, green the young crops and green grass; yellow piles of wheat, white the water, blue the sky, orange the golden rays of light and red the sun.

扫码收听 音频内容

裕固族
Yugur

　　裕固族源出唐代游牧在鄂尔浑河流域的回鹘人，目前主要聚居在中国甘肃省肃南裕固族自治县和酒泉黄泥堡地区。裕固族最有特色的服饰是女子的"头面"。这是两条从背后搭在胸前的镶饰宽带，遍镶银箔珠穗，长可及地，重达数斤。关于"头面"有一个古老的传说：很久以前，有两座大山，一座叫白山，一座叫黑山。白山下住的是裕固族的白头目，黑山下住的是外族的黑头目。为纪念沙场战死的白头目娘子军萨尔玛珂，姑娘出嫁时必须戴头面。头面上镶嵌的红色珠子，表示萨尔玛珂的乳房；白色海贝表示萨尔玛珂的白骨；帽尖上缀的红缨穗，表示萨尔玛珂头顶的鲜血。

　　The Yugur people originated from the Uyghurs, nomads of the Orkhon basin during the Tang dynasty. Currently, the Yugur mainly live in Sunan and Jiuquan of Gansu province. The most distinctive costume of the Yugur are the women's head ornaments called *Toumian*. These are two wide bands that hang over the chest from the back. The bands, which are heavy and long enough to reach the ground, are inlaid with silver foil and beaded tassels. There is an old legend about this ornament. A long time ago, there were two mountains, one of which was called White Mountain and the other Black Mountain. The White chieftain of the Yugur people lived under the White Mountain, while under the Black Mountain was the Black chieftain of a foreign tribe. To remember Saermake, the White chieftain's wife who died on the battlefield, girls must wear this ornament when they are married. The red beads inlaid on the ornament represent Saermake's breasts; the white seashells represent her bones and the red

tassel adorning the tip of the hat represents the fresh blood on Saermake's head.

扇面毡帽配长袍，展示男性魅力
Fan-shaped Hat with Long Robes

裕固族男子一般穿大襟长袍，多用布、绸、缎或白褐子（一种手工粗布）缝制，长及脚面，系红色或蓝色腰带。夏秋季戴金边毡帽，后帽檐卷起，后高前低，呈扇形，帽镶黑边，帽顶为蓝缎制作，上有用金线织成的圆形或八角形图案。

Yugur men generally wear long robes, which are usually made of cloth, silk, satin or white coarse cloth. These robes are long enough to reach the tops of the feet and tied red or blue belts. A felt hat with a gold edge and rolled up back brim is worn in summer and autumn. The brim at the back is high while that at the front is low. It is fan-shaped. The hat is set with black edges and the top of the hat is made of blue satin with a circular or octangular pattern woven in gold thread.

"头面"与红缨毡帽——夸张的立体之美
Toumian with Red Hat

裕固族的女帽，只有已婚的妇女才能戴，形似礼帽，顶部缀饰下垂的红线穗子。而上文提到过的"头面"堪称精致的民间工艺品，它以红布、青布或红色香牛皮做底，再用珍珠玛瑙穿缀成各种图案，分别系在发辫上。

Yugur women's hats can only be worn by married women. They resemble a top hat and are decorated with hanging red tassels on the top. In addition, the *Toumian* mentioned above is a fine folk craft. The base is made of red cloth, green cloth, or fragrant red cowhide, and then various patterns are stitched and plaited using pearl agate. Finally, tie it to women's braids.

扫码收听 音频内容

东乡族
Dongxiang

东乡族主要分布在中国西北的甘肃省、宁夏回族自治区境内洮河以西、大夏河以东的山麓地带，其中半数以上聚居在东乡族自治县，部分散居在新疆。现今的东乡族与回族、维吾尔族、蒙古族等多民族混居，其服饰特点则取众家之长：如白帽、白衫、白坎肩，三角绣花巾、冬日里穿的大皮袄等，都受到兄弟民族的影响，但女子的帽箍与坎肩，却有自己的独特之处。

The Dongxiang people are mostly distributed in northwestern China, including Gansu and Ningxia. More than half of them live in Dongxiang autonomous county and some are scattered in Xinjiang. Today's Dongxiang people live alongside other ethnic groups including the Hui, Uyghur and Mongol and the characteristics of Dongxiang clothing are based on the strengths of each group. For example, hats, shirts and sleeveless jackets, all of which are white; embroidered triangular cloths and winter fur-lined jackets were all influenced by the other ethnic groups, however the ribbon hats and sleeveless jackets worn by women have their unique features.

西北特色的撞色之美
The Beauty of Contrast in the Northwest

名为"过美"的绣花裙，是东乡族女性的婚礼服。此裙多为绿、黑、浅蓝三色，裙下摆绣有约17厘米宽的花边。包括前后开衩的长袍以及镶有假袖的斜襟上衣。有的短衣袖口上，缀以很多假袖，以显示富有。

The embroidered skirts called *guomei* are worn by Dongxiang women on their wedding day. This dress, which is mostly green, black and pale blue, is embroidered along the lower hem with lace of about 17 centimetres in width and includes a robe that is slit in the front and back and a slanting jacket with imitation sleeves. Some of these tops are decorated with multi-layered imitation sleeves to show wealth.

女孩们独特的小圆帽
Unique Little Rounded Hats for Girls

如今东乡女孩们喜戴的褶子帽，颇有特色，帽子的圆顶多用蓝色或绿色，帽檐用绿色丝绸卷起褶皱作为大圆边，右侧戴绢花及丝穗。

The pleated hats worn by Dongxiang girls today are quite distinctive. The dome of the hat is often blue or green in colour and green silk is rolled up to form a large rounded brim. Silk flowers and silk tassels are worn on right side of the hat.

扫码收听 音频内容

保安族
Bonan

目前主要聚居在中国甘肃省临夏县大河家，因原居住地在青海同仁隆务河边的"保安三庄"（保安城、下庄、尕沙日）而得名。保安族服饰早期受蒙古族影响，男女冬季均穿长皮袍，戴各式皮帽；夏秋穿夹袍，戴白羊毛毡制作的喇叭形高筒帽，系各色丝绸腰带，并佩有小装饰物。清同治年间，保安族迁到甘肃大河家，以后的较长时间里，因与回、东乡、汉等民族来往密切，以及生产活动的需要，其服饰有了比较明显的变化。

At present, the Bonan people mainly live in Gansu province and they are named as such after the place they originally came from, *Bonan three villages*. Bonan costumes were influenced by the Mongolians in the early days. Both men and women wore long leather gowns and various kinds of leather hat in winter. In summer and autumn, they wore a lined robe, a tall white woollen and felt hat in the shape of a funnel; various colours of silk belts and small ornaments. During the Tongzhi period of the Qing dynasty, Bonan moved to Dahejia of Gansu where their clothing changed significantly as a result of their close contact with other ethnic groups including the Hui, Dongxiang and Han, as well as their manufacturing needs.

刀背藏身，硬朗之美
Waist Knife

腰刀是保安族男子重要佩饰，始于清代后期，技术成熟于民国时期。品种繁多，以双刀和双垒刀最著名。刀体经过反复锻打，异常锋利。刀面

还往往刻上星星、梅花、龙、手等精美图案。

The waist knife is an important accessory for male Bonans, beginning in the late Qing dynasty, its technology was developed during the period of Minguo. Of the many varieties, the most famous kinds are the dual blade and the double-base knives. The blades are particularly sharp after repeated forging and are often engraved with exquisite patterns of stars, plum blossoms, dragons and hands.

绉绉帽——文艺百搭，气质首选
Female Hat

年轻的保安族妇女爱戴用绿色丝绸制作的"绉绉帽"，帽子为平顶，帽边用同色或其他色绸缎堆绉而成，上面绣花或缀以珠串、绢花、璎珞等，有的在帽子左边吊一束穗子，穗上还串一个绣荷花包。

Young Bonan women love wearing crepe hats made of green silk. The hat has a flat top and its wrinkled sides are made with satin of the same or different colour. It is embroidered or adorned with strings of beads, silk flowers and festoons. Some people hang tassels which hold an embroidered lotus bag on the left side of the hat.

藏族
Tibetan

跨境民族，青藏高原的原住民。中国境内主要分布在西藏自治区、青海省和四川省西部，云南迪庆、甘肃甘南等地区。粗犷豪放的藏族服饰，与生活在世界屋脊的这个民族的气质相当契合。藏族服装令人难忘的是藏女飞舞的长袖和霓虹般的"氆氇"围裙，是康巴汉子半袒的皮袍……其实，只要深入藏区，你会发觉，不同地区的藏装，均有形制各异的服饰创意，同样让人目不暇接。

The Tibetans are the indigenous people of the Qinghai-Tibet Plateau. The majority in China live in Tibet, Qinghai and western Sichuan, Yunnan diqing, southern Gansu. The rugged Tibetan costume is fitting for the people that live on the Roof of the World. What are most unforgettable about Tibetan dress are the neon-coloured *Pulu* aprons and the dancing long sleeves worn by women, together with the half-covering leather robes worn by men. However, there is much regional variance in the styles of Tibetan costumes, and each is a feast for the eyes.

冷峻、野性与温柔并存的男性魅力
Masculine Charm

康巴藏族男子一般将衣袍下摆提升至膝盖以上，脱两袖扎于腰际，腰间除火链等佩物外，一把长刀十分醒目。头上或盘起饰有红丝线的独辫，或戴呢礼帽；耳戴镶松石的单挂耳环。

Tibetan men usually position the lower hem of their robe above the knee and tie the sleeves around the waist. In addition to fire chains and other waist pendants, a long and eye-catching knife is also carried.

They either wear a single braid decorated with red silk or a heavy woollen hat. They also wear a single turquoise earring.

历久弥新的经典条纹围裙
Classic Striped Aprons

藏族妇女标志性的花条纹围裙"帮典"，一般用藏毛呢"氆氇"制成，以红、绿、蓝、黄、白五色为基调，色彩有时多达十四至二十种，条纹越细，越素雅，越显示穿着者的身份高贵。

The iconic *bangdian*, which is a colourful striped apron worn by Tibetan women, is usually made of *pulu*, a type of Tibetan woollen cloth, and consists of five key tones: red, green, blue, yellow and white. There are sometimes as many as 14 to 20 colours used in this garment. Thinner stripes convey greater elegance, which in turn reflects the noble class of the bearer.

五彩斑斓的饰品让美丽层层绽放
Colourful Accessories

拉萨妇女饰品多数集中于头部和胸部，头部的羊角形或三角形"巴珠冠"十分讲究，在藤条编成的帽架上饰满珍珠、珊瑚、玉、松石等珠宝，尊贵又精致。

The accessories worn by the women of Lhasa mainly adorn the head and chest. The goat-horn-shaped or triangular pearl crowns are both tasteful and of quality, and this is visible from the abundance of pearls, coral, jade and turquoise that exquisitely decorate the rattan hatstand.

扫码收听 音频内容

珞巴族
Lhoba

　　跨境民族，中国境内主要分布在西藏自治区东南边陲的珞渝地区，与藏族关系密切。在喜马拉雅山南麓，如果你遇到一位戴熊皮帽的汉子，那一定是珞巴族猎手。珞巴族的熊皮帽专挑长毛黑熊的皮制作，戴时长毛披于后颈，如同长发，加上腰间必佩的长刀，形如武士甲衣的宽肩毡制无袖长袍和长筒皮靴，显得彪悍勇武。女子服饰以繁饰为特色，风格原始中透着华丽。至近代，随着珞巴人同藏族联系愈加密切，输入羊毛，引种棉花，使衣料材质和形式更加多元。

　　The Lhoba people are closely conn- ected with Tibetans and those living in China are mainly distributed in Luoyu area at the southeast border of Tibet. If you meet a man wearing a bearskin hat in the southern foothills of the Himalayas, he must be a Lhoba hunter. The Lhoba bearskin hat is specifically made from black bear fur and, when worn, the bear fur covers the back of the neck just like long hair would. A long knife is invariably carried on the waist and coupled with a felt sleeveless broad-shouldered robe, which is shaped like samurai armour, and long leather boots, making the wearer look intrepid and valiant. Female clothing is characterised by many decorations. The style is authentic and resplendent. In modern times, with the Lhoba people becoming closer to Tibetans, they introduced wool and cotton to diversify the materials and forms of their clothing.

戴熊皮帽的汉子雄壮威武
Bearskin Hat
珞巴族服饰最引人注目的便是熊

皮帽。珞巴人善狩猎，有些健壮男子一生用于狩猎的时间长达四十年。头戴熊皮帽，肩挎毒箭筒，腰系长刀，手持强弓是珞巴猎人的标准装扮。

The most noticeable item of Lhoba clothing is the bearskin hat. The Lhoba are good at hunting and some of the fittest men can spend up to 40 years practising this activity. Wearing a bearskin hat, shouldering a poison arrow quiver, carrying a long knife on the waist and holding a strong bow is the Lhoba hunter's standard outfit.

传奇贝壳腰饰——从大海中来
Shell Waist Ornaments

珞巴族女子传统腰饰，珞巴语称"乌克"，将小海贝串成球状花形，每两串系在一根牛皮条上，共10串，由5根纵向牛皮条分别系于横向的牛皮绳上，垂下来像一簇花丛，十分夺目。

The traditional waist ornaments worn by Lhoba women consist of shells. Small seashells are linked and shaped into balls, and each pair of strings are tied to a cowhide strip. Five cowhide strips are connected vertically to five horizontal strips so as to produce a very eye-catching bunch of flowers.

扫码收听
音频内容

门巴族
Moinba

　　跨境民族，中国境内主要分布在西藏自治区东南部的门隅和墨脱地区，与藏族长期友好往来，呈大聚居、小杂居状态。门巴族常用酒和歌来抒发美好而淳朴的情感，同时，在民族服饰上也显示出对美的追求与向往。生活在藏南山谷中的门巴族，服饰以大襟绲边皮毛袍和毡制无袖长袍为主，也爱穿绣花毡靴，佩珠链，戴大耳环；特色服饰为男女均戴的小帽"拨尔甲"。

　　The Moinba people living in China are mainly distributed in Menyu and Medog southeast regions of Tibet. They have had longterm friendly exchanges with Tibetans, with whom they live together in large and small settlements. The Moinba people often use alcohol and song to voice their beautiful and honest emotions. They have also shown their pursuit and longing for beauty in their costume. The people living in the valleys of southern Tibet wear a large-breasted fur-lined robe and felt sleeveless robe. They also prefer to wear embroidered felt boots, pearl chains and big earrings. Their special clothing is the small cap of men and women.

时髦的缺口小帽
Fashinable Cap

　　门隅地区的门巴族，男女均戴一种叫作"拨尔甲"的小帽。多为深色帽顶，橙色帽檐，帽檐前留有一个楔形缺口，男子戴时缺口在右眼上方，缺口自然敞开；女子戴时缺口朝后，用彩色布绲边，别具特色。

　　Male and female Moinba living in Menyu area wear a small cap which mostly has a dark top and orange brim. There is a

wedge-shaped gap in the front part of the brim and, when this hat is worn by men, the naturally opened gap is positioned above the right eye, while when it is worn by women, the notch faces backwards and a colourful trim is added, making it unique.

文成公主之爱
The Legend of Princess Wencheng

上门隅的邦金、勒布一带的门巴族妇女习惯在背上披一张牛犊皮或羊皮。关于这一装束，民间有种传说：相传唐代文成公主进藏时，曾披一张毛皮以避妖邪，后将此皮赠予门巴族妇女，于是沿袭下来成为本民族特色服饰。

It is customary for Moinba women living around Bangjin and Loeb in Shangmenyu to wear calfskin or sheepskin on their backs. There is a folk legend about this costume. It is said that princess of Wencheng in Tang dynasty wore fur when entering Tibet so as to repel evil spirits; later, she gave this item to the Moinba women, who continued wearing it, in doing so, made it into a characteristic item of Moinba clothing.

扫码收听音频内容

羌族
Qiang

主要分布在中国四川阿坝藏族自治州的茂县、汶川县、理县、黑水县、松潘县，甘孜藏族自治州的丹巴县以及绵阳地区的北川县等。作为古老的"西戎牧羊人"，羌族的历史至少可以上溯到商周时代。崇尚白色的羌族，服饰也以白为贵：白袍、白裤、白头巾、白羊毛坎肩等。传说在清朝咸丰年间，黑虎乡一带时常遭到外敌侵犯，寨子里一位机智勇敢的羌民率领乡亲保卫家园，人称"黑虎将军"，英雄牺牲后，大家都穿戴白色服饰，作为对他的纪念。随着时间推移，羌族服饰也逐渐趋向色彩缤纷。但传统的白色，仍在关键的部位显示着尊贵的地位。

The Qiang people are mostly distributed in Aba,Ganzi of Sichuan. As the ancient *Xirong Shepherds*, the Qiang history can be traced back to at least the Shang or Zhou dynasties. Seeing the Qiang advocate white, this colour is valued in their clothing: white robes, white trousers, white headscarves and white woollen waistcoats. According to legend, during the Xianfeng period of the Qing dynasty, the Black Tiger township was frequently invaded by foreign enemies. A resourceful and brave Qiang citizen of the stockaded village known by the title *Black Tiger General* led his fellow villagers to defend their home. After the hero's sacrifice, everyone wore white clothes in memory of him. With the passage of time, Qiang clothing gradually become more colourful. However, the traditional white colour still shows its noble status in key areas of clothing.

脚上有朵花做的云
Cloud Shoes
羌族男女都穿自制的"云云鞋"，

此鞋因鞋面常绣有彩色云纹而得名。云云鞋鞋尖微翘，形同小船，除云纹外，鞋梁两边还贴绣有虎头纹、灵芝纹、水波纹等，显示了羌族妇女的高超手工艺。

Both Qiang men and women wear self-made *cloud shoes*, which take their name from the colourful clouds often embroidered on the uppers. The tips of these shoes are slightly raised, like a boat, and, apart from the clouds, embroidered patterns of tiger heads, fungi and water ripples are stuck on both sides of the shoe. This shows the superb handicraft of Qiang women.

古老习俗的温暖力量
The Warmth of Ancient Customs

羌族男女都习惯包头帕。男子缠青、白色头帕或戴狐皮帽。缠头帕时在顶部留出一小撮头发，据说进入山林万一被蛇咬或受弩伤，可剪下一绺头发，烧成灰撒在伤口上，以防病毒侵害身体。

Qiang men and women have the habit of wearing turbans. Men wear green or white turbans or fox skin hats. A small tuft of hair is left exposed when wrapping the turban. It is said that if ever someone is bitten by a snake or struck by a crossbow while entering a mountain forest, a lock of hair can be cut and its burnt ashes will prevent the venom from invading the body if spread on a wound.

扫码收听音频内容

彝族
Yi

中国人口较多的少数民族之一，主要分布在云南、四川、贵州、广西四省区的高原与沿海丘陵之间，凉山彝族自治州是中国最大的彝族聚居区。彝族尚黑，服饰一般以黑色或青色为基调，衬以红、黄等色，分别象征刚强尊贵、热烈和善良。传说历史上曾有"六祖分支"的故事，所以服饰也有六大类型：凉山型、乌蒙山型、楚雄型、滇西型、红河型、滇中及滇东南型。不过，由于彝族分布太广，支系太多，服饰的变体估计不下百种。

The Yi is one of China's populous ethnic groups and they mostly live between the highlands or coastal hills of Yunnan, Sichuan, Guizhou and Guangxi while China's largest Yi settlement is in Liangshan Yi Autonomous Prefecture. Black is valued among the Yi people and black or cyan is used as the base colour of clothing which is lined with either red or yellow as a symbol of strength, dignity, warmth and friendliness. Legend has it that there are six types of Yi ancestral branches, so there are six types of clothing: Liangshan, Wumengshan, Chuxiong, Dianxi, Honghe, Dianzhong and Diandongnan, all of which are named after places or regions. However, due to the wide distribution and many branches of the Yi, it is estimated that clothing has no fewer than 100 variations.

霸气英雄髻——凝聚男性魅力
Hairstyle of the Yi Men

四川凉山地区，男子头部以缠巾为饰。青年男子用长巾裹成细如竹笋的尖锥状头饰，长约30厘米，斜插额上，显得威武雄壮，俗称"英雄髻"。

Men in Liangshan wear a turban. Young men wrap their turban so as to form a tip of about 30 centimetres which is as pointed and fine as a bamboo shoot, and the turban itself is positioned diagonally across the forehead. This garment is commonly referred to as a hero bun because it making the wearer look powerful and full of grandeur.

百褶长裙——定义专属时尚
Long Pleated Skirts

百褶裙是彝族妇女传统下装，其中以凉山布拖县最为典型。当地彝族的百褶裙用纯羊毛织成，质地柔软。中部为红色或红黄相间的长筒状；下段拼接蓝、红、白、黑相间的细条纹，走起路来皱褶摇曳，轻盈飘洒，分外俏丽。

The pleated skirt is the traditional bottom garment of Yi women and its most representative version comes from Butuo county of Liangshan. The skirt has a soft texture because it is made of pure wool. Red and yellow are used in the middle and the bottom is stitched with a sequence of blue, red, white and black. This skirt is particularly pretty when the steps of women make it gracefully float and sway.

扫码收听 音频内容

傣族
Dai

　　跨境民族，中国境内主要分布在云南省的怒江、澜沧江、元江、金沙江流域，近一半人口聚居于西双版纳和德宏。以秀媚来概括傣族女子服饰，是比较恰当的，但不同地域支系的傣族服饰又各有特点。西双版纳傣族均穿无领紧身短衣和长筒裙，极精炼地显现出傣女娇好的身材；红河的花腰傣在短衣长裙上镶绣许多银饰和花边，美得令人眩目；滇西北的傣族，代表服饰是大袖短衣和火草布筒裙，腰带扎出鱼鳍的感觉；而滇南的黑傣，则穿大袖半臂衣，又是另一种格调。图中呈现的是花腰傣服饰。

　　The Dai people living in China are mostly distributed in Yunnan province and about half the population are settled in Xishuangbanna and Dehong. People describe Dai girls' clothes as pretty and elegant.It should be noted that Dai costumes coming from different regional branches have their own characteristics. Dai girls living in Xishuangbanna wear short and tight collarless tops and long straight skirts which reveal their beautiful figures. The Huayao Dai of Honghe embroider dazzling silver adornments and lace to their short tops and long skirts. The costume of the Dai people from northwest Yunnan is represented by large-sleeved short tops, a fire-grass cloth dress and a tight belt. The Hei Dai of south Yunnan wear a large-sleeved top that only covers half their arms. The picture shows the Huayao Dai costume.

媲美 UFO 的鬼马 "鸡枞帽"
Unique Dai Hat

宛如飞碟落在了头上，这款帽檐微微上翘形成碟状的笠帽，是花腰傣女孩的日常装束。形状除了像飞碟，也很像云南的一种鸡枞菌。戴上它，衬得女孩们个个古灵精怪，是人群中最亮眼的风景线。

As part of their everyday attire, Huayao Dai girls wear a hat made of bamboo which has a slight raised brim to form the shape of a saucer. This out of the ordinary hat makes each and every girl stand out in a crowd.

傣族女孩的万能宝袋
A Secret Pocket Under the Dress

花腰傣女子会将筒裙的下摆外翻折转，上提牵拉形成凹槽状的兜兜，这个巧妙玲珑的口袋可以存放零食杂物，十分别致实用。

The Huayao Dai girls pull the hems of their skirts upwards to form the groove of an ingenious pocket which can be used for storing snacks and other bits and bobs.

扫码收听音频内容

阿昌族
Achang

　　跨境民族，中国境内主要聚居在云南省德宏傣族景颇族自治州陇川县户撒乡，梁河县囊宋乡、九保乡，其余分布于潞西、盈江、腾冲、龙陵、云龙等县。阿昌族服饰的差异性与生产方式的差异息息相关。梁河县的阿昌族以农业为主，妇女们一般按照传统技术制作自己的服饰，民族特点能够得到较为完整的保留。而陇川县的阿昌族手工业比较发达，服饰制作受到了其他民族的影响，他们利用当地银器手工业的优势，在服饰上添加了许多银饰，并改变了服饰的色彩、材质和造型。

　　The Achang People mainly live in Dehong of Yunnan. The differences in Achang costumes is closely connected to the the variation of modes of production. The Achang living in Lianghe county are agricultural; women usually make their own clothes according to traditional techniques, meaning ethnic characteristics can be completely preserved. By contrast, the handicraft industry of the Achang living in Longchuan county is relatively developed and their clothes production has been influenced by other ethnic groups. By taking advantage of their local silverware handicraft industry, they were able to add many silver ornaments to their clothing and change the colours, materials and shapes.

阿昌男孩利刃出鞘
The Achang Boy's Unsheathed Blade

　　传说明洪武年间沐英三征麓川，在户撒建立沐庄，并留下军匠打造武器。这些军匠逐渐融入了当地阿昌族

的生活，之后刀具便成为阿昌族男性的随身物品。直至今日，过节的时候阿昌小伙都会佩戴阿昌刀，显得英俊威武。

Legend has it that during the Hongwu period of the Ming dynasty, an army came and left military craftsman to build weapons. These craftsman gradually integrated into the local Achang lifestyle and knives became an belonging Achang men carried on themselves. Until today, Achang lads will carry a Achang knife during festivals, making them look powerful and mighty.

立下赫赫战功的包头
Useful Turban

阿昌妇女们的包头高度在各民族中夺得榜首。关于包头有一个动人的传说，一次战争时一位聪慧的母亲想出了一个绝妙的主意，让前方战士们戴上高高的包头，误导敌人把箭射向包头，不仅挽救了不少战士的生命，还用这种方式获得了敌方的箭镞。为纪念这位妇女和立下战功的包头，阿昌族女性延续了戴高包头的习俗。

The headdress worn by Achang women are the tallest among all the ethnic groups. There is a story about this item. During a war, a mother came up with a brilliant idea which was to let frontline soldiers wear a tall headdress; by wearing this, the enemy soldiers were tricked into shooting their arrows at the headdress, which not only spared many lives, it also allowed Achang soldiers to take possession of the cluster of enemy arrows. Achang women continue the custom of wearing this item as a way of commemorating the woman and the headdress that proved itself outstanding in battle.

景颇族
Jingpo

　　跨境山地民族，中国境内主要分布在云南德宏傣族景颇族自治州的陇川、潞西等地。当景颇族妇女在传统节日"目瑙节"的集体舞中，一起抖动双肩，让满肩银泡和银坠随鼓声刷刷作响时，每个人都会被那样的气氛感染。黑衣、银泡、红色筒裙、宽边红包头，是景颇妇女最典型的装束。传说女祖先本为龙女，银泡即是龙鳞，她与太阳之子结合，繁衍了刚直豪放的景颇族。

　　The Jingpo people living in China are mostly distributed in Dehong, Yunnan. During the traditional *Munao Festival* collective dance, an infectious atmosphere is created when silver balls and pendants attached to the shoulders of Jingpo women chime to the beating of drums. The most typical dress for Jingpo women consists of black clothes, silver balls, a red skirt and a tall red headscarf. According to legend, their ancestor was a female dragon and the silver balls represent the dragon's scales. This dragon married with the son of the sun to produce the honourable and bold Jingpo people.

龙鳞化身的银泡披肩
The Silver-ball Shawl

　　景颇族女性身上的银泡披肩，相传是来源于其始祖宁贯娃娶龙女为妻繁衍后代，龙女化身为人之后龙鳞就变为银泡状的披肩。为表示对始祖母的崇拜，求得平安顺遂，景颇族妇女就有了用银泡制作披肩的传统。

　　The silver-ball shawls worn by Jingpo women are said to have originated from their ancestor, Ningguanwa, who married

a female dragon to breed descendants. The female dragon became human and her scales turned into a silver ball-like shawl. In order to worship their first maternal ancestor and seek peace and success, Jingpo women have a tradition of making shawls with silver balls.

舞蹈担当"瑙双"四哥
Dancing Clothes

"瑙双服"是景颇族最古老的祭祀服装，十分有特色。"瑙双"指祭祀时的领舞者，一般是四位男性。走在前面的两位头戴藤条编制的犀鸟长喙椭圆形帽，身穿红色长衫；后两位穿戴相似，长衫改为黄色，头部少了犀鸟长喙。

The very distinctive *Naoshuang clothing* is the Jingpo's oldest sacrificial ceremonial costume. *Naoshuang* refers to the dancers during sacrifice ceremony, which generally consists of four males. The two male dancers in front wear red gowns and oval-shaped plaited rattan hats that are decorated with long hornbill beaks. The two behind are similarly dressed, however their gowns are yellow and without hornbill beaks.

扫码收听
音频内容

德昂族
De'ang

中缅交界地区的山地跨境民族，中国境内主要分布在云南省德宏、保山、临沧等地。穿短衣长筒裙的德昂妇女，与众不同的是她们腰间的藤圈。传说女人原为天女，漫天飞行，男人奈何不得，只好用藤圈将她们套了下来。德昂族无论男女，都喜欢在包头、肩、头、襟口、挎包等处装饰很多以红色为主色调的绒球。

The De'ang people are mountain people living between China and Myanmar, and those living in China are mostly distributed in Dehong, Baoshan and Lincang in Yunnan. De'ang women wear short tops and long skirts, but what is different is the waist rings they wear. Legend has it that women were once celestial beings, flying in the sky; men had no option but to use rings to pull them down. Both men and women decorate their headwear, shoulders, top lapels and shoulder bags with red pompoms.

传统又前卫的爱情信物
Lover's Waistband

腰箍既是年龄的表示，也是美的标志，更是青年男女的爱情信物。在求爱和交往过程中，男孩往往要精心制作有花纹图案的腰箍送给心上人。

The waistband is not only a symbol of age, it is also a mark of beauty and token of love between young men and women. During courtship, men often have to make a patterned waistband for their lover.

军装制服范儿的时尚双排扣
Double Buttons

作为德昂族妇女的胸饰，和西方

军队制服中的大衣双排扣有异曲同工之妙，排扣上雕刻有精细的花纹，通常为鹿、鸟、花草及几何纹，由本民族手艺高超的银匠手工制作。

De'ang women wear chest ornaments which resemble the two rows of buttons on a double-breasted army uniform jacket. The buttons are hand-engraved by superb silversmiths with fine patterns, usually those of deer, birds, flowers and geometric shapes.

在胸前绽放的五彩绒球
Colourful Pompoms

不同颜色的绒球星星点点相互呼应，琳琅满目，使黑色上衣立马变得光彩夺目，是德昂族最为骄傲与喜爱的装饰。

The tiny pompoms of differing colours work well together as glittering jewels that add brilliance to the black tops; this is a decoration of which De'ang people are both fond and proud.

扫码收听
音频内容

怒族
Nu

　　跨境民族，中国境内主要分布在云南省怒江傈僳族自治州的泸水、福贡、贡山、兰坪，以及迪庆藏族自治州的维西和西藏察隅等地。怒族是一个善于博采众长的民族，他们的服饰反映了与周围民族的密切关系：喜欢织与独龙毯相似的条纹麻布，用两块麻布围成裙装；男子喜佩刀持弩，显得豪迈英武。他们喜戴贝饰。传说怒族的先民们迁徙时追赶西行的太阳，来到怒江石月亮山，为了识别族群、铭记祖先，妇女们把这段记载着本民族历史印迹的真实生活融入服饰中，制作出寓意深远的贝壳珠帽。

　　The Nu people of China mainly live in Nujiang, Diqing of Yunnan, with some living in Tibet. The Nu ethnic group are a group that is good at learning from others and their costume represents their close relationship with nearby ethnic groups. They like to knit striped linen cloth which is similar with Drung blankets, and use two pieces of linen cloths to make a skirt. Men carry a knife or crossbow to show their strength. They also like to wear shell decorations. Legend has it that the ancestor of the Nu people chased the sun towards the west to Stone Moon Mountain by the Nu River. In order to commemorate their ancestor, women have included this history into their clothing in the form of a meaningful shell and pearl hat.

能当钱用的珠珠帽
Bead Hats as Currency

　　海贝曾是古老的流通货币，为了方便保管，怒族女性和藏族一样，喜欢把贝壳串起来戴在头上。怒族女子

头饰一般为珊瑚、海贝等串成的半月形珠帽。

Shell is an ancient currency , in order to keep it safe, Nu women like to string them together and wear them on their heads like Tibetans. Nu women usually use coral and seashells to make crescent moon-shaped bead hats.

贝壳项链——身份与财富的象征
Shell Necklace — Status and Wealth

怒族女性常见的胸饰，一般在胸前佩戴一枚直径约 5 厘米的圆形大贝壳，另有长短不一的数串彩珠分别挂在胸前或斜挂在身上。

A common ornament worn by Nu women is a big round shell measuring five centimetres in diameter. They also wear lots of colourful bead necklaces of varying lengths in front of the chest or diagonally along the body.

扫码收听
音频内容

独龙族
Drung

　　跨境民族，中国境内主要聚居于云南省贡山县独龙江流域的河谷地带。独龙族是中国人口较少的民族。自古以来，独龙妇女习惯用腰机将麻线织成条形毯，缀连成片，斜披裹于身上。现在，独龙族男女老少已穿上各式各样的现代服装，但是保暖而洒脱的"独龙毯"，仍为他们喜爱。

　　The Drung people living in China are mainly concentrated in the Dulong River Valley in Gongshan county of Yunnan province. This ethic group has a relatively small population in China. In ancient times, Drung women were used to using a waist loom to weave thread into blanket strips which were embroidered into one piece and then wrapped diagonally around the body. While the Drung people wear all sorts of modern clothing nowadays, they still love their warm and unrestrained *Drung blanket*.

一毯多用，镇族之宝
A Multipurpose Blanket: A Treasure to the People

　　独龙毯是独龙族的标志性服饰，白天为衣，夜晚为被。独龙妇女精于织麻，从种麻剥皮、漂晒纺线开始，全是手工操作，制作成的独龙毯较宽大，男女老少均能穿着，穿法简单快捷，披落自如随意。

　　The Drung blanket, which acts as clothing in the daytime and a quilt at night, is iconic to the Drung costume. Drung women are adept at weaving linen and the whole process is manual, from the planting and peeling of flax, to the bleaching and drying of spinning yarn. The blankets are

made wide so that they can be worn by men, women or children. Putting on the blanket is simple and fast and it can be draped in any way.

傈僳族
Lisu

跨境民族，中国境内多数聚居于滇西、滇西北的怒江、澜沧江和金沙江两岸的河谷山坡地带。傈僳族服饰因居地不同而差异很大，因而又有以服饰颜色命名支系的习惯。白傈僳穿麻布长褂或长裙，以白贝为饰；黑傈僳多穿黑衣黑裙，以青布包头；花傈僳已成泛指，各地都有独特的装束：永胜县的傈僳族"花"在裙上，采用彩色布条拼缀裙身；华坪县的"花"在衣上，衣服上有长条状的彩色拼布绣；保山一带则全身皆花，让人眼花缭乱。

Many Lisu people living in China inhabit west and northwest of Yunnan. Lisu costumes vary greatly from region to region and it became customary to name the branches after the colour of their costumes. Bai (White) Lisu people wear linen gowns or long skirts which are decorated with white shells. Hei (Black) Lisu people wear black clothes (tops and skirts) and wrap their heads in cyan cloth. The term Hua lisu has been used in a general sense to mean there are unique costumes everywhere. The Lisu people of Yongsheng county use strips of coloured cloth pieced together on their dress, while the Lisu of Huaping county embroider long strips of colourful patchwork on their tops. Finally, those living in Baoshan area wear clothes that are colourful all over.

傈僳女孩们的贴心小褂
Tops Worn by Lisu Girls

妇女的短衣，傈僳语称为"皮度"，长及腰间，没有纽扣，平时衣襟敞开穿着，配色对比强烈，简洁干练。

The short tops worn by Lisu women, called *pidu* in Lisu language, are buttonless, usually worn open and drop down to the waist. Colours are tastefully matched and contrast is strong.

篱笆花包头的使命
Fence Flower Turban

傈僳男子的包头称"篱笆花包头"，是男孩们展示英俊的标志，这种包头不仅能起到护发和保暖的作用，还能抵挡外力的侵袭。

The turban worn by Lisu men is called a *fence flower turban* and is a sign that boys use to show they are both handsome and spirited. This turban not only protects the hair and provides warmth, it also acts as a helmet protecting the wearer from harm.

用生命来保护的"龙尾巴"
The *Dragon's Tail*

由于傈僳族是从中国西北部迁徙而来，他们普遍认为自己是龙的传人，在服饰上保留了飘带的设计，对其珍视有加，所以坐下前先掀起飘带，不能压到"龙尾巴"。

Owing to the fact that the Lisu people migrated from northwest China, they generally regard themselves as the descendants of the dragon. They continue the design of including a streamer on their clothes, because this aspect is treasured even greater today, they lift it up before sitting down so as not to sit on the *dragon's tail*(the streamer).

纳西族
Naxi

大部分纳西族居住在滇西北的丽江市，其余分布在云南其他县市和四川盐源、盐边、木里等县，也有少数分布在西藏芒康县。丽江纳西妇女服饰中最突出的要数七星羊皮披肩，已被公认为纳西族的典型服饰，一说"披星戴月"，人们马上想到纳西族。但纳西族服饰还有其他特色，比如香格里拉白地一带的纳西族妇女，穿开口长褂配百褶裙，头上装饰彩线和金属饰品，身上斜披长毛白山羊皮，似乎是延续了《后汉书》等史书中记载的"牦牛种"（纳西族在汉代的称谓）的服饰习俗。

Most Naxi people live in Lijiang of northwestern Yunnan and the rest are dispersed in other counties and cities in Yunnan, Yanyuan Yanbian and Muli in Sichuan, Mangkang in Tibet. The seven-star sheepskin shawl, which is an item worn by Naxi women of Lijiang, is recognised as typical Naxi clothing. People will immediately think of the Naxi when they hear someone say "wear the stars and the moon". However, Naxi costumes have other characteristics, for example, Naxi women from the region of Baidi in Shangri-La wear an open gown with a pleated skirt and their heads are decorated with coloured threads and metal adornments. They also wear a long-hair white goatskin diagonally on the body, which seems to be a continuation of the clothing customs of the *Yak Species*, which was the appellation of the Naxi during the Han dynasty, as recorded in historical books including the Hou Han Shu.

颜值担当"七星披肩"
The Exquisite *Seven-star Shawl*

"肩挑日月，背负七星"，象征纳西妇女披星戴月、吃苦耐劳的精神。七星披肩不仅颜值高，还有保暖防寒和保护背部等功能。

"Shouldering the sun and moon and bearing the seven stars" symbolises the hardworking and enduring spirit of Naxi women. Not only is the seven-star shawl exquisite, it also functions to keep the wearer warm and protect the back.

戴在头上的小太阳
The Small Suns Worn on the Head

纳西女孩的头饰中通常有 10 至 12 个刻有太阳图案的圆形银牌，环绕点缀着盘在头顶的发辫，简洁明艳，是纳西盛装中最靓丽的风景线。

Naxi girls usually have 10 to 12 round silver medals which are engraved with images of the sun on their headdress. The plaits dotted around the top of the head are simple and bright, making it the most beautiful aspect of Naxi dress.

普米族
Pumi

主要聚居在云南省怒江州的兰坪县、丽江市的宁蒗县、玉龙县和迪庆州的维西县。其余分布在云南的云县、凤庆、中甸以及四川的木里、盐源、九龙等地，与汉、白、纳西、藏等民族交错杂居。崇拜白额虎的普米族，服饰也有尚白之俗。兰坪等地的普米族习惯穿白大襟短衣，披白羊皮坎肩，裹白绑腿；宁蒗一带的普米族妇女则爱在大襟上衣上再披一块白羊皮，下穿百褶长裙。普米族男子服饰多为大襟立领布衣，外套皮袍，也常将皮袍褪至腰间穿着。

The Pumi people mostly live in Nujiang, Lijiang, Diqing of Yunnan but others are distributed in Sichuan, where they are mixed with Han, Bai, Naxi and Tibetans. The Pumi worship the white-forehead tiger and have the custom of wearing white clothes. The Pumi people of Lanping usually wear short right-buttoned tops, white sheepskin sleeveless jackets, and white leg wrappings. Pumi women from Ninglang like to wear white sheepskin on top of their jacket, and a long pleated skirt. Men's clothing mostly consist of straight-collared clothes and a fur-lined jacket which often falls down to the waist.

高端定制羊皮披肩
Superior Shawls

普米族的披肩用上乘白色山羊皮制成，上面装饰毛呢、皮草、翡翠、海贝、玛瑙等等，高端大气上档次，保暖又奢华。

Pumi shawls are made of superior white goatskin and decorated with wool,

fur, jade, seashells and agate, making it look high-end, warm and luxurious.

围在腰间的彩虹
Rainbow Belt

普米族腰带是自织自染的精美工艺品，分麻织、毛织两种。男女均用，系在长衫之外，因编织过程倾注了时间和心血，姑娘们常将亲手织的腰带作为信物赠给男友。

Pumi belts are an exquisite homespun and dyed handicraft. The belts are made of hemp or wool. They are worn over gowns by men and women alike. Because much time and effort is required in knitting this garment, a girl will often give a hand-knitted belt to her boyfriend as a keepsake.

白族
Bai

云南独有民族之一，有学者认为其族源可追溯到石器时代。主要聚居在大理白族自治州的平坝和低山丘陵地带。生活在苍山脚下、洱海之滨的白族人服饰以白为美。白族姑娘喜穿白色或浅彩色上衣，长发独辫，头顶绣花头帕。白族男子多包白布头帕，着白色对襟上衣，束腰带，外罩黑蓝色坎肩，下穿宽松裤，裹有装饰了边纹的裹腿。白族人心灵手巧，工艺精湛，他们的服饰生动地诠释了"风花雪月"四个字本身。

As one of ethnic groups that are unique to Yunnan, some scholars believe the Bai origin can be traced back to the Stone Age. The clothing of the Bai people who live at the foot of Cangshan Mountain and on the shore of Erhai Lake is white because this colour is regarded as beautiful. The Bai girls like to wear white or pale colour tops, plait their hair and don embroidered scarfs. The Bai men often wear turbans made of white cloth, white Chinese-style buttoned jackets, belts, outer-garments consisting of a bluish-black sleeveless jacket, roomy trousers and decorated leg wrappings. The Bai people are deft with their hands, producing exquisite craftsmanship, allowing their clothes to vividly annotate the four words of wind, flower, snow and moon.

金花头饰里的小心机
Headbands Made by the Golden Flowers

白族姑娘头顶装饰有由方形刺绣、扎染方巾折叠而成的绣花头帕，秀发盘压于头帕下，用红色头绳缠紧，左

耳侧垂下头帕丝线。精巧绝伦，金花们的手艺不言而喻。

An embroidered headband made of folded tie-dyed square cloth decorates the heads of the Bai girls. Coiled hair is pressed under the headband and is wrapped in red plaiting string and some of the headband hangs next to the left ear. The ingenious workmanship of the Bai girls (who are also known as golden flowers) is most evident.

玩转盖头的潮男阿鹏
Apeng's Headdress

白族男性盖头是战乱时期简化婚礼程序的产物，用意一是在仪式上遮羞，二是辟邪。阿鹏的盖头由两片方巾叠加而呈八角形。其实，白族男孩俊朗帅气，"女粉"众多，展示都还嫌少，遮羞那倒不必。

The headdresses worn by the Bai boys, which resulted from simplified wartime wedding procedures, have two purposes that are to hide shyness during the ceremony and to ward off evil spirits. The headdress of the Apeng (a name used to refer to the Bai boys) is an octagon formed by layering two pieces of square cloth. In fact, because these handsome Bai boys have many female fans, there is no need to be shy.

扫码收听
音频内容

佤族
Va

　　跨境民族，中国境内主要聚居于澜沧江、萨尔温江之间，怒山山脉南段一带。云南的沧源和西盟是佤族的主要聚居县。佤族的先民是濮人，史称"望蛮"。由于分布区域的不同，佤族服饰也表现出不同的特征，而以沧源和西盟的最具代表性。佤族崇拜红色和黑色，服饰多数以黑为主，红为辅，保留着古老的山地民族特色。妇女喜戴大耳筒，宽手镯等，标志着她们粗犷、豪放的坚强性格。男子则喜文身，其中多数在胸脯刺牛头、手腕刺鸟、腿上刺山林图案；外出时常挎花布袋，佩戴长刀或火枪，给人以雄壮威武之感。

　　The Va people living in China are mostly settled between the Lancang and Salween rivers and south of the Nushan Mountains. The main counties of Va settlement in Yunnan are Cangyuan and Ximeng. The ancestors of the Va people were the Pu people, who were known in history as *wangman*. Due to being settled in different areas, the Va costumes vary in their characteristics, with the Cangyuan and Ximeng types being the most representative. The Va people worship red and black, therefore most of their costumes are black and supplemented with red color, retaining the characteristics of ancient mountain people. Women wear large earrings and wide bracelets to mark their strong rugged and bold personality. Men on the other hand have tattoos; many have a bull's head tattooed on their chest, a bird on their wrist and mountain forests on their legs. When going out, men often carry a cotton embroidered bag over their shoulders and carry a sword or musket,

which gives a sense of power and grandeur.

举世无双的长发伴侣
The Long–Haired Va Women

伍族女性以长发披肩为美，头戴月亮形银质发箍。发箍一般用银子或者铝片弯折制作，两端有螺旋状带扣。正上方雕刻有十字花、牛角等图案，十分显眼。

Va women regard long hair as beautiful and they wear a silver headband in the shape of a crescent moon. The headband is made of silver or aluminium and has screw buckles at both ends. Conspicuous patterns including the Chinese character for ten and bull horns are engraved on the headband.

手臂上的耀眼光芒
Dazzling Radiance on the Arm

伍族女子皮肤多为小麦色，上臂裸露，臂饰是必不可少的装饰品。常见臂饰有圆筒式的银质臂箍，箍身较宽，箍面雕刻有花草和麦穗图案，阳光下闪闪发光。也有用藤篾编制的臂箍。

The skin of most Va women is tanned, their upper arms are exposed and arm accessories are essential decorations. A ubiquitous accessory is the cylindrical silver armband that sparkles in sunlight and which is wide and decorated with engravings of flowers, plants and wheat ears. There are also armbands that are made of platted cane.

扫码收听
音频内容

布朗族
Blang

　　跨境民族，中国境内分布在云南省西部和西南部澜沧江中下游西侧的山岳地带，主要聚居在西双版纳傣族自治州勐海县布朗、打洛等山区。布朗族服饰色彩绚丽、绣艺精湛、缝工讲究、极富特色，传统服装从种植大树棉花到纺织品成形，无一不是手工制作。青婆罗缎、桐华布、绵绢、彩帛、文绣等为早期布朗族服饰提供了丰厚的物质基础。布朗女子常上着窄袖紧身短衣，下穿双层条纹筒裙，勾勒出她们婀娜的身姿。

　　The Blang people living in China are mostly distributed in western and southwestern Yunnan Province. Blang clothing is very characteristic. It is colourful and exquisitely embroidered and sewn. Traditional clothes are fully handmade, from growing large cotton trees to making textiles. Some clothing techniques provided a rich material basis for early Blang costumes. Blang women often wear short tight-fitting tops that have narrow sleeves and a two-layered striped skirt, all of which outlines their graceful bearing.

布朗女性的终极护发神器
The Ultimate Blang Hair Care Product

　　布朗族女子头上习惯缠布包头，未婚女性包头上一般以彩色线球、纽扣、彩色珠子或银泡装饰。已婚女性的包头通常无装饰，保暖美观，是随身携带的护发神器。

　　Blang women cover their heads in cloth and unmarried women usually decorate the cloth with coloured thread

balls, buttons, coloured beads and silver balls. Those who are married usually leave their headwear undecorated. This item, which protects the hair, is both warm and beautiful.

黑红蓝三色创世神话
Black, Red and Blue: Three Colours of the Creation Myth

布朗族男女的上衣袖口一般都有黑红蓝三道布条装饰，黑代表大地，红代表火塘，蓝代表天空。布朗族认为天地万物有灵，天地造就了先祖，火塘繁衍着后代，所谓"日子不断，火塘不灭"。

The cuffs worn by Blang men and women are usually decorated with black, red and blue stripes. Black represents the earth, red represents the fire pit and blue represents the sky. The Blang people believe everything in this world has a soul, the heaven and the earth created the ancestors and the fire reproduces the descendants, namely, *the days are unceasing and the fire is inextinguishable.*

哈尼族
Hani

　　跨境民族，中国境内主要分布在元江和澜沧江之间，聚居于云南的红河、江城、墨江及新平、镇沅等县。哈尼服饰以奇著称，其中以爱伲一支最突出，妇女喜穿露脐短胸甲、百褶超短裙，外套绣花外衣，胸甲饰满银泡，头饰更是繁复，缀满银泡、红缨、牙骨和花草贝壳。由于支系不同，服饰更是千差万别，很难一语述尽。比如上图呈现的哈尼族奕车支系妇女，喜穿无领开襟敞怀短衣和紧身超短裤，头披尖形披肩帽。

　　The Hani people living in China are mostly situated between the Yuan River and Lancang River with many concentrated in Yunnan. Hani clothes are famous for their unusualness. Among them, the Aini people are the most prominent. Women wear short corsets, pleated mini-skirts and embroidered jackets. The corsets are decorated with silver balls as is a complicated headdress. In addition to the balls, the headdress is also decorated with red tassels, bones, flowers and shells. Costumes differ greatly because of the various branches, making it difficult to describe them all. The example picture shows what the women belonging to the Hani Yiche branch wear, namely a collarless open short top, tight shorts and a pointed cape that is worn over the hair.

C位出道的时尚"龟服"
Fashionable Female Clothing

　　奕车女孩的上衣，分雀朗（外衣）、雀巴（衬衣）、雀帕（背心）三种，用自制自染的靛青土布为料，款式为无领开襟高腰，搭配紧身超短裤，时尚度爆表。

The tops of the Yiche girls are divided into three types including outerwear (*quelang*), shirt (*queba*) and vest (*quepa*) and are made with self-dyed indigo handwoven cloth to make collarless open high-waist tops and matching tight-fitting shorts.

人脚一双的木拖鞋
A Pair of Clogs

将两节与脚大小相近的木头砍成小板凳状，做成一对夹角木拖鞋，一般高 4 厘米。不仅是哈尼女孩脚上的增高神器，还是夏日必备单品。谁说时尚的完成度只能靠脸？

Wooden angular slippers are made by cutting foot-length wood into small blocks of about 4 centimetres in height. Not only is this way of increasing height, it is also a must-have during the summer.

拉祜族
Lahu

　　跨境民族，中国境内主要分布在澜沧江西岸，北起临沧、耿马，南至澜沧、孟连等县。有着"猎虎民族"之称的拉祜族，服饰上却绝无剑拔弩张之势。男子普遍身着黑色布衣，头缠包头，形制和色调都很朴实。女子服饰也以黑色为主，拉祜西支系穿短衣长筒裙，饰红色布条；拉祜纳支系穿斜襟长衫，喜用三角和方形彩色布条缀饰衣裙下摆两侧。

　　The Lahu people living in China are distributed on the west side of Lancang River and nearby area. While the Lahu are known as the *tiger-hunter ethnic group*, you would not tell from the clothes they wear. A man is generally dressed in plain shapes and tones consisting of black clothes and a turban. The clothing worn by women is also mostly black, with the people of Lahu Xi branch wearing short tops and long skirts which are decorated with red strips. The people of Lahu Na branch wear a sloping lapel long gown which has colourful triangular or square patches decorating the hems.

南北交融的华丽长"尾巴"
The Gorgeous *Long Tail* that Blends the North and South

　　拉祜族支系拉祜纳的妇女喜穿长衣，相较于其他西南少数民族的短上衣，反而更具有北方民族的服饰特点。长衣制作工艺复杂，下摆开叉，衣服边缘点缀有不同颜色几何形的拼贴，远看像是彩色的尾巴一般，摇曳生姿。

　　Lahu Na women wear long clothes. Compared to the short clothes worn by

other ethnic groups in the southwest, they have more characteristics resembling the dress of northern ethnic groups. The production process of the long dress is complicated: the hem is split and the edges are decorated with a collage of different colours and geometric shapes. From afar, it looks like a colourful swaying tail.

基诺族
Jino

　　云南独有民族之一，主要聚居于云南省西双版纳傣族自治州景洪市的基诺山基诺乡。基诺族男子一般外穿无领对襟长袖麻布短衣，衣后背饰有黑底红、橘、黄、白四色的绣花图案"孔明印"。下着麻布长裤，前裆垂有一片兜裆布条，它反映了基诺先民早期着装的遗俗。基诺族妇女穿麻线织成的色彩艳丽的衣裙，称为"彩虹衣"，胸前戴精美图案的挑花胸兜，胸兜上不可缺少的纹样是折线纹。这大约是葫芦藤的变形，代表着创世传说。

　　The Jino people are one of several ethnic groups that are unique to Yunnan province and they mainly live in Xishuangbanna Dai Autonomous Prefecture. Jino men usually wear long-sleeved collarless linen tops, the backs of which are black with a pattern embroidered in red, orange, yellow and white known as *Kongming print*; and linen trousers which have a strip of cloth in front of the crotch that reflects the early clothing customs of the Jino forebears. Jino women wear *rainbow dresses*, which are colourful dresses that are woven in flaxen thread, and a stomacher that is elegantly patterned, cross-stitched and includes an indispensable crease pattern.

绣在背上的"孔明印"
Kongming Print

　　关于基诺族的族源有一个传说，说他们的祖先是三国时代跟随诸葛亮由北南征而来，所以男女服装的背部都绣有一个特殊图案，状如"孔明印"。

　　There is a legend about the origin of the Jino which says their ancestors

followed Zhuge Liang from the north during the Three Kingdoms period; this is why the backs of the clothes worn by men and women are embroidered with special prints like the *Kongming print*. Kongming is a courtesy name for Zhuge Liang.

创世之母竖立的时尚标杆
Legend of a Female Ancestor

传说基诺族的女始祖阿嫫腰白从水中诞生时，头戴白色尖顶帽，于是基诺族女人们受创世女神影响，都爱戴白色的尖顶"泳帽"，穿美丽的"比基尼"（状如肚兜），践行着"内衣外穿"的新时尚。

According to legend, the female Jino ancestor A'moyaobai wore a white pointed hat when she was born in water. Influenced by the genesis goddess, Jino women started wearing a white pointed *swimming hat* and a beautiful *bikini* (which is in fact a garment covering the chest and abdomen), thereby practising a new fashion of "wearing underwear on the outside."

收放自如的尾巴
Jino Men's Tail

基诺族男子裤装后裆垂有一条长长的砍刀布尾巴，传说尾巴摇曳之时，人就会像无人机一般飞翔（仅是传说），闲时将尾巴系于腰间，方便劳作。

There is a long knife shaped cloth tail at the back of the trousers worn by Jino men. According to legend, there is a humorous story that says, when the tail is shaken, people will fly like drones. When they do not use this tail, it is fastened to the waist, doing this helps them work.

扫码收听 音频内容

蒙古族
Mongol

素有"马背上的民族"之称的蒙古族，从北到南都有他们的足迹，也是一个跨境民族。中国境内主要分布在内蒙古自治区、东北三省、新疆、河北、青海，其余散布于河南、四川、贵州、北京和云南等地。在内蒙古，无论牧区农区，人们日常均穿袍衫，着皮靴，扎头巾，唯盛装时各地区差异较大：有的戴黑丝绒绣龙平顶圆帽；有的用松耳石、玛瑙、珊瑚、珍珠做成发箍或胸饰；有的戴直径一米多的白布斗笠……而在云南，蒙古族衣装却已变短，其样式也是蒙、彝风格混合了。

Having the name of *the people on horseback*, the footmarks of the Mongol can be found from north to south and spanning borders. The majority of Chinese Mongols live in Inner Mongolia, Heilongjiang, Jilin, Liaoning, Xinjiang, Hebei, Qinghai and other areas ,like Henan, Beijing, Yunnan, etc. Irrespective of whether they live on Inner Mongolian pastureland or farmland, they typically wear robes, leather boots and bandanas, while their splendid attire varies from region to region: some wear black velvet flat hats with embroidered dragons; some use turquoise, agate, coral and pearls to make headbands or chest ornaments; some wear white hats that are a metre in diameter, in Yunnan, the clothing worn has become shorter in length and resembles a mixture of Mongol and Yi styles.

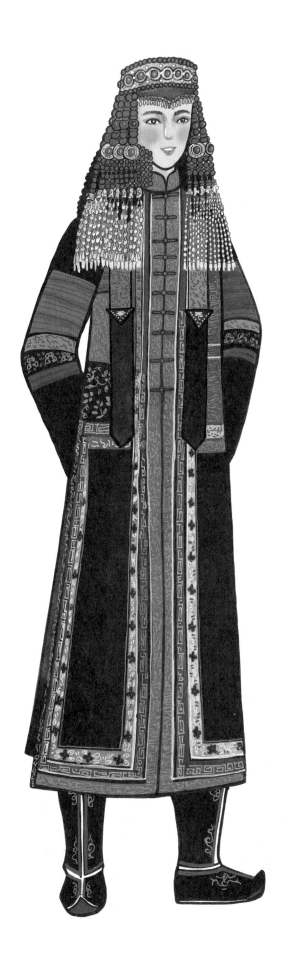

四季不离身的蒙古袍
Mongolian Robes—Suitable for All Seasons

　　蒙古族无论男女老幼，一年四季均喜欢穿蒙古袍。春、秋穿夹袍，夏季穿单袍，冬季穿皮袍或棉袍。冬季在远行时有的还穿两件皮袍，牧民的皮袍多为羊皮材质。

　　Males and females of all ages usually wear Mongolian robes all year round. Lined robes are worn in spring and autumn, unlined robes are worn in summer and robes made of leather or cotton are worn in winter. People occasionally wear two leather robes when travelling long distances in winter and the leather robes worn by herdsmen are mostly made of sheepskin.

头饰——美与财富的象征
Headdress—Symbol of Beauty and Wealth

　　头饰是蒙古族贵族女性服饰中最绚丽的部分，一般由银质发夹和额箍组成，整体呈牛角或羊角形，上面镶嵌绿宝石与红珊瑚，天然形成红白绿的主色调，与黑色发辫交相辉映。

　　The Mongol headdress, which is one of the most magnificent items worn by Mongolian noblewomen, is usually composed of silver hairpins and hoops that form the shape of a cow or sheep horn, and is inlaid with emeralds and red coral. The naturally-formed red, green and white tones are complemented by black plaited hair.

扫码收听
音频内容

土家族
Tujia

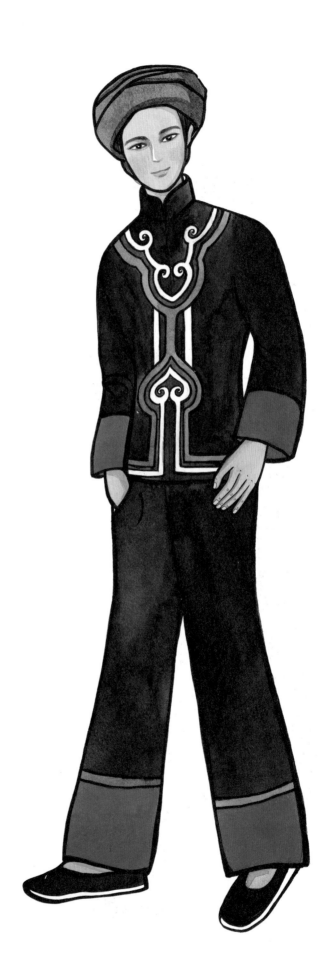

    主要分布在湖南湘西、张家界，湖北鄂西、宜昌，四川黔江以及贵州铜仁等地。土家人善于用织锦和刺绣美化自己的服饰。年轻姑娘腰膝间必有一块色彩艳丽的织锦"西兰卡普"，俗称"土锦"或"斑布"。作为土家族的特产，出嫁时的嫁妆和跳"摆手舞"时的披甲也都是土锦，土锦织得好说明这家姑娘聪慧勤劳。土家族男子对襟上衣的缘边也很有特点，为宽缘的云纹。男子穿裤，女子则裙裤皆宜。

    The Tujia people are mostly settled in Hunan, Hubei, Sichuan and Guizhou Provinces. They are good at using brocade and embroidery to adorn their clothing. A bright and beautiful brocade known as a *xilankapu*, also commonly know as *Tu brocade* or *spotted cloth*, decorates the clothing worn by young girls. As a special Tujia product, the dowry given at the time of marriage and the armour worn during the hand-waving dance is Tu brocade. Well-woven Tu brocade shows a girl is both smart and hardworking. The edge of the Chinese-style buttoned jacket worn by the Tujia men is also very characteristic because of the cloud patterns along its wide margins. Men wear trousers, women wear either skirts or trousers.

风度翩翩的云纹上衣
Cloud Patterned Tops
    土家族成年男子一般头包青丝帕或青布帕，上身穿对襟布衫，对襟上衣的缘边为宽缘的云纹，下装多为青、蓝色裤子，白布裤腰，裤脚肥大而较短，脚穿青布面白底鞋。

Tujia men usually wear silk or cloth cyan turbans, buttoned shirts which have cloud patterns decorating its broad edges, cyan or blue roomy and rather short trousers with white waistbands and cyan shoes which have white soles.

裹在发辫中的美丽心机
Tujia Hairstyle

土家族未婚女青年不包头帕，头上梳一条长辫，将红头绳编于发辫之中、缠于头顶或拖在身后，头上插花。盛装时，胸前挂一大串由银制珠链串成的胸饰，行走时叮当作响。

Rather than wear a headscarf, an unmarried Tujia woman plait her hair with red string which is then either coiled into a bun or left to dangle behind her back. Having done that, flowers are added on the head. During special occasions, women wear chest ornaments made of silver beads and chains that jingles when they walk.

扫码收听
音频内容

苗族
Miao

　　跨境民族，人口较多，中国境内主要分布于贵州、湖南、湖北、四川、云南、广西、海南等省区。以蜡染和刺绣闻名的苗族，自古便喜"五色斑衣"。苗族服饰，不仅种类多、形式美，而且文化内涵非常丰富。苗女的衣裙上，形象地记述着苗族的古老历史和传统文化，人们可以据此了解许多悲欢离合的故事，是一部随身携带、辗转千年的象形史书。

　　The Miao has a fairly large population. In China, most live in Guizhou, Hunan, Hubei, Sichuan, Yunnan, Guangxi and Hainan. The Miao people is famed for batik, which is the use of wax to print colour on cloth, and embroidery. Since the past they have worn variegated clothes known as five-colour speckled clothes. There are many types of beautifully-styled Miao clothing which are rich in cultural meaning. Ancient Miao history and traditional culture is vividly recorded on the dresses worn by Miao women, enabling people to uncover numerous joyful and sorrowful tales, just like a thousand-year-old pictographic history pocket book.

雉尾冠——一枝独秀，做人群中最帅的仔
Pheasant–Tail Crown

　　在跳花节上，苗族男青年们身着织锦多层领上衣，披绣有几何纹样的黄金色半臂式披肩，头戴雉尾冠，手持葫芦笙，以此装束表演箐鸡舞，博取姑娘欢心。

　　During the Dancing Festival, young Miao men wear a brocaded multi-layered

collared jacket, a golden half-arm shawl with geometric patterns and a pheasant-tail crown while holding a Hulusheng(a reed windpipe musical instrument). This dress is worn during the the Chicken Dance, which is performed to attract the attention of girls.

牛角形银帽——精美繁复的工艺品

Horn–Shaped Silver Headwear

银帽已成为黔东南苗族侗族自治州的苗族少女标配，当地苗族以此作为美丽、高贵、富有的标志。苗族崇拜水牛，以牛为美，所以银帽均仿牛角形状制作，立体繁复，是全手工打造的精美工艺品。

Silver headwear is worn by Miao girls living in southeastern Guizhou, and locals regard this as a symbol of beauty, high-class and wealth. Water buffalo are worshiped by The Miao people and are seen as beautiful, that is why the silver headwear is always horn-shaped. This complex headwear is fully crafted by hand.

侗族
Dong

主要分布在中国贵州省的黔东南苗族侗族自治州、铜仁市，此外湖南、广西，湖北等地也有侗族聚居。侗族姑娘喜穿百褶短裙，无领上衣长而敞襟，露出胸前的绣花肚兜，因无领并梳高髻而愈显秀美的脖颈上挂着几层银项圈，恰到好处地将开襟上衣和胸兜连为一体，与深色的衣裳形成对比。短裙下是绑腿和绣花鞋，秀丽中不乏潇洒利落。

The Dong people mostly live in Qian dongnan and Tongren of Guizhou province, other provinces where they reside include Hunan, Guangxi and Hubei. The Dong girls wear a short pleated skirt and a long collarless top with an open lapel which reveals an embroidered undergarment. In addition, layered silver neckbands are worn, the beauty and elegance of which is emphasised by the collarless top and the fact the girls wear their hair in a bun. The ensemble of collarless top and embroidered undergarment are perfectly integrated and in contrast with dark clothes. Leg wrappings and embroidered shoes are worn under the skirt.

金属之美，做朋克女孩
The Beauty of Metal

侗族女孩节庆时喜戴银项圈，一般是四个，由小到大，霸气十足，胸前通常还挂有两条银链及银珠。平时便装则较朴素，少有装饰。

The assertive-looking neckbands are worn during the festivals and usually consist of four rings, from small to big, and two silver chains with silver beads. Everyday clothing is simple and unadorned.

闪闪惹人爱的侗族包头
The Shining and Lovely Dong Headband

    黔东南的侗族男子盛装时包较大的紫色闪光侗布头帕，帕端绣有红、绿色锯齿纹样，包头式样颇有讲究，节日期间还会戴银帽或插羽毛为饰。

    On special occasions, the Dong men living in Qian dongnan wear splendid and large sparkling purple turbans which have red and green zigzagging patterns. During festivals, silver hats and feathers are also worn as ornaments.

布依族
Bouyei

　　跨境民族，中国境内主要分布在贵州、云南、四川等省，其中大部分聚居在贵州省的黔南和黔西南。清新淡雅的布依族服饰，以青、白等素色为基调。妇女服饰式样较多，各地多有不同，传统的布依族妇女服装以花边大领衣和百褶长裙为主要特征。男子服饰较为简练，中青年男子包头帕，上穿对襟短衣，一般内衣为白色，外衣为青色或蓝色，下着长裤。另有部分布依族妇女穿精美的翘头绣花鞋，戴笋壳青布做的头饰"假壳"，很有特色。布依族服饰集蜡染、扎染、挑花、织锦、刺绣等多种工艺于一身。

　　The Bouyei people living in China are mostly situated in Guizhou, Yunnan and Sichuan, with many living in Qiannan and Qianxinan prefectures Guizhou. The Bouyei customs are based on colours such as cyan and white. There are many styles of women's clothing which vary from place to place, however traditional clothing is mainly characterised by lace collars and long pleated skirts. In contrast, male clothing is simpler. Young and middle-aged men wear kerchiefs, short buttoned jackets and long trousers. Colour-wise, their undergarments are usually white while the overgarments are blue or cyan. Another group of Bouyei women wear exquisite pointed embroidered shoes and unique-looking replica shell head ornaments, which are made of bamboo shoot shells and green cloth. Bouyei costumes combine various techniques such as batik, tie-dying, cross-sticking, brocading and embroidery.

二次元上的圆形律动
The Pattern in Bouyei Clothes

贵州镇宁扁担山的布依族女孩服饰极具特色，上衣是青布小袖窄腰大开襟衣，袖口和衣摆均镶有7厘米宽的织锦花边，袖筒上有较宽的三节绣花镶边。袖子中部纹样为六个圆并列，每个大圆又由七个小圆花组成，充满结构感。

The clothing worn by Bouyei girls of Zhenning in Guizhou is particularly distinctive. A cyan narrow-sleeved open-buttoned jacket that is tight-fitting around the waist is worn. Its cuffs and hem have a seven-centimetre-wide brocaded lace edge and the sleeves have a wide three-sectioned embroidered trim. The well-structured pattern in the middle of the sleeves consists of six circles and each large circle consists of seven smaller circles.

百褶裙上的舞蹈节奏
Long Pleated Bouyei Skirts

布依女孩多穿长及脚背的百褶长裙。上半段为青布，中间为蜡染花布，蓝多白少，下段是蜡染花布，白多蓝少，细密有致，形成色彩鲜明度渐变的节奏感，裙上蜡染纹样为银杏果和蕨菜花，既可爱又活泼。

Bouyei girls wear long pleated skirts that fall down to the top of their feet. This fine and detailed garment has three sections: the upper part of this skirt consists of cyan cloth and the middle and bottom both consist of batik printed sections. Blue is more common than white in the middle section while the inverse is true of the bottom. This forms a rhythmic sense of colour gradation. The batik patterns onto the dress consist of images of ginkgo fruit and fiddlehead ferns, making the dress charming and full of life.

扫码收听音频内容

仡佬族
Gelo

主要聚居于贵州北部的道真、务川、遵义、仁怀、金沙及西部的织金、黔西、六枝、大方、关岭、安顺等地，有一部分散居在云南及广西。仡佬族因服饰上的差异，民间有"青仡佬""花仡佬""红仡佬""白仡佬"和"披袍仡佬"等称谓。女子一般穿长袖衬衣，外套蜡染和彩绣半袖外衣、对襟坎肩，下着百褶长裙并加小围腰，头巾罩在发髻上，余幅后垂至肩背；男子包头，穿对襟密织上衣、长裤。按传统习俗男未娶者以金鸡羽毛为头饰，女未嫁者则饰以海螺。

The Gelo people mainly live in northern Guizhou province, with some living in Yunnan and Guangxi. Due to differences in clothing, the Gelo ethnic group have various names such as *Green Gelo, Flower Gelo, Red Gelo, White Gelo* and *Shawl Gelo*. Women usually wear long-sleeved shirts, batik coats and embroidered half-sleeved outerwear and sleeveless jackets. For the bottom part, they wear pleated skirts and small aprons; Some of the headscarf worn covers a hair bun, with the rest hanging over the shoulders. Men also cover their heads, wear tight-fitting top and long trousers. According to traditions, unmarried men use rooster feathers as head decorations while unmarried women use sea snail shells as decorations.

穿梭云海间的"筘云子"
Reed Clouds

仡佬族服饰一般镶饰云彩纹或浪花纹的宽边，称为"筘云子"，花

边上用花线绣"吊线"。如遇隆重场合，还加穿绣有多种花式图案的"提裤"。年轻女性多穿百褶裙，上衣外面套坎肩。

Gelo costumes are generally decorated with wide borders of cloud or wave patterns, which are known as *reed clouds*. They embroider colourful *hanging thread* onto the lace edges. During important occasions, women wear *tiku* trousers, which have many flower patterns. Young girls generally wear pleated dresses and sleeveless jackets.

仫佬族
Mulam

主要聚居在中国广西壮族自治区西北山区的罗城仫佬族自治县，少数散居在广西的忻城、柳城、都安、融安等地和贵州的麻江、凯里等地。仫佬族的日常服装追求简朴的风格和精巧的搭配，以实用、轻便为着装原则。善耕作的仫佬族，男子戴黑色或深色包头，穿对襟上衣，系布腰带；女子穿大襟无领或浅领上衣，系绣花胸兜，上衣边缘用很宽的色布或花布绲边，下装多为裤子，脚穿绣花鞋，梳辫或盘髻，喜戴银质首饰。

The Mulam people mostly live in the mountains northwest of Guangxi. In their everyday clothes, they pursue simple styles and exquisite matching, which follow the principle of practicality and lightness. The Mulam people are good at farming. Men wear a black or dark turban, a jacket and a cloth belt. Women wear collarless or shallow-collared tops, which have wide embroidered edges made of colour or print cloth, embroidered stomachers, trousers and embroidered shoes. They wear their hair plaited, or in a bun, and add silver jewellery.

衣袖上跳动的音符
Special Sleeves

仫佬族妇女一般穿右衽、窄袖、有领的素色上衣；胸前襟边镶浅蓝色布，其上有两条细线作装饰；两边袖口各绣一条花边，这是整件上衣中最为醒目之处，使朴素的装束有了点睛之笔。

Mulam women usually wear plain collared tops which have a right-sided

lapel and narrow sleeves. The chest is lined with a lapel of light-blue cloth which is decorated with two thin lines. The sleeves have embroidered cuffs, which being the most eye-catching part of the garment, help make the simple clothes more elaborate.

水族
Shui

　　主要分布在贵州和广西交界的都柳江和龙江上游一带，大部分聚居于贵州省三都水族自治县，以及荔波、独山、都匀等地，还有少量散居在广西和云南。水族服饰以黑为主，黑包头、宽袖对襟黑衣、黑裙或黑裤，妇女胸前系一黑色围裙。都是黑色，却不沉闷，原因在于胸前和袖口的彩绒花边，它们在黑底中点缀了一抹亮色，勾勒出庄重素雅的形象。据说她们裙边裤角的花边，还有驱蛇的作用呢。

　　The Shui ethnic group is mainly spread across Guizhou and Guangxi, most live in Sandu Shui Autonomous county of Guizhou province. The Shui people mostly use black in their clothes; they wrap their heads in black, wear wide-sleeved buttoned jackets, black dresses and black trousers. Women also wear a black apron. Though what they wear is black, it is not morbid, because colourful laced velvet, which decorates their chests and cuffs, adds a touch of vibrancy that covers the black base and outlines a solemn yet elegant image. According to legend, the lace adorning the hem of their shirts has the power to repel snakes.

银饰与黑衣的反差之美
The Contrasting Beauty of Silverware and Black Clothes

　　水族妇女日常多用黑色或方格头巾包头，但盛装时却截然不同：在发髻前戴有"出"字形的银冠，银冠下端挂满各式银花、银片和银链，胸前垂一片月牙形的压领，下吊银链、银铃，构成了上半身的主要装饰。

Shui women frequently wrap their heads in black or chequered scarves, however their festive dress is completely different: they wear a silver crown on top of hair that has been tied into buns, below the crown, hangs an assortment of silver flowers, leaves and chains. Over their chests they wear a collar that is in the shape of a crescent moon and which has silver chains and bells hanging down. This is the main way they adorn their upper bodies.

一身青衣的大气沉稳
Dark Garments Which Add Composure

水族男子一般上身外穿及腰的青色对襟衣，下配大裤脚长裤，头包青色或黑色头帕，有时头帕外露出缨须加以点缀。

Shui men wear tight-fitting cyan shirts and baggy long trousers. On their heads they tie a cyan or black turban which has tassels as a decoration.

扫码收听
音频内容

满族
Manchu

　　中国人口较多的少数民族之一，分布于全国各地，以辽宁、吉林、黑龙江和内蒙古、北京、河北等地为主。满族服饰因清王朝近三百年的统治而推向极致。传统满族男服较典型的为长衫马褂；女服则为宽身长袍加坎肩，头梳"二把头"，戴黑色扇状头冠，脚穿高跟马蹄底或花盆底绣花鞋。旗袍早期较宽大，后来渐紧窄，两侧开衩较高。

　　The Manchu People are a fairly populous group in China and are spread out throughout the country, with many living in Liaoning, Jilin, Heilongjiang, Inner Mongolia, Beijing and Hebei. The Manchu costumes were pushed on the people during the near 300-years' rule of the Qing dynasty. The typical traditional male costume consisted of a long gown worn under a mandarin jacket, while women wore a broad gown under a sleeveless jacket, hair was combed in two parts under a black fan-shaped crown and high-heel or pot-bottomed embroidered shoes were worn. The early qipao was initially rather wide-fitting before gradually becoming narrow with high slits on both sides.

国民之服——旗袍
The Qipao—China's National Costume

　　旗袍满语称为"衣介"，男人着旗袍便于鞍马骑射，女人的旗袍在衣襟、领口、袖边等处镶饰花绦或彩牙边。清代至民国年间，无论汉满、男女、城乡、贫富，一般人都有一件旗袍，只是发展到近代，旗袍的样式有了较

多改变。

The qipao was know in Manchu as *Yijie*. The qipao was convenient for men who practised pommel horsemanship and archery, while the version worn by women was decorated around the placket, neckline and cuffs with decorative flowers or colourful moon-shaped patterns. From the Qing dynasty to the Minguo period, the average person owned at least one qipao, irrespective of whether they were Han or Manchu, male or female, resident of a city or the countryside and rich or poor; it was only in modern times that the qipao changed a lot.

长衫马褂的满族阿哥
The Mandarin Jacket

马褂最初是骑马时穿的外褂，既适于骑射的需要，又防风御寒，使用广泛。满族进关以后，虽不再过射猎生活，但上自君王，下至平民，都习惯了穿马褂。

The mandarin jacket was originally a garment worn for horseback riding as it met the requirements of riders and archers in addition to protecting the wearer from the wind and cold. Over time, the Manchus from all strata continued to wear mandarin jackets even though they no longer lived the life of hunter with bow and arrow.

达斡尔族
Daur

　　草原民族，主要分布在中国内蒙古自治区莫力达瓦旗、鄂温克旗，黑龙江省齐齐哈尔市郊梅里斯达以及新疆塔城等地，呈大分散、小聚居状态。善骑射的达斡尔族男子穿前襟下摆正中开衩的大襟长袍，以便乘骑；善猎者喜戴动物头形的皮帽，佩短刀，穿高筒皮靴。妇女穿右襟长袍，腰系金银线腰带，或在袍衫之外套穿深色绣花坎肩，扎头巾。在皮制品上刺绣，是达斡尔族妇女擅长的传统工艺。

　　The Daur people are a grassland people and are mainly distributed in small settlements including the Morin Dawa and the Ewenki banners of Inner Mongolia; the Meilisi Daur district of Qiqihar in Heilongjiang province and Tacheng in Xinjiang. Daur men are talented horseback archers and the long robe they wear, which has a slit from the middle of garment to the hem, helps them with this activity. Hunters like to wear leather hats resembling animal heads, tall leather boots and a knife. Women wear long robes that are fastened to the right and either belts made of gold and silver thread or dark embroidered waistcoats. Women also wear headscarves. Leather goods are embroidered by Daur women, a traditional craft they do well.

小小烟荷包——达斡尔专属时尚单品
Daur Smoking Purse

　　达斡尔族传统烟具，也可作装饰品。分男女多种样式，一般用绸缎等布料缝制，形状各异，葫芦形的最为常见，用各色丝线绣花草、鸟虫或人

物图案于荷包上，工艺精细，实用美观。

The traditional Daur smoking purse can also be used as a decoration. There are various styles for men and women and they are generally made using satin and of various shapes, with the gourd-shaped ones being most common. The patterns of grass, birds, insects and human figures are embroidered on the purses with various colours of silk string. Not only are these purses finely crafted, they are also both practical and pleasing to behold.

扫码收听音频内容

鄂温克族
Ewenki

　　跨境民族，主要居住于俄罗斯西伯利亚以及中国的内蒙古和黑龙江两省区，以善养驯鹿著称。由于居地分散，又多与其他兄弟民族杂处，不同的经济、文化背景及生存环境，决定了各地鄂温克人不同的服饰特点。内蒙古的鄂温克族，男女都穿绣有很宽花边的大襟长袍，但女子的长袍却别有韵致：下摆宽大多褶，静若百褶裙，动似喇叭口。她们还常戴一种结有红缨的阔边毡帽，帽檐上翻。无论男女，都扎宽腰带，穿皮靴。

　　The Ewenki people mainly live in Siberia of Russia and Inner Mongolia and Heilongjiang provinces of China. They are known for being adept at breeding reindeer. Many Ewenki people are mixed with other ethnicities because of their scattered settlements and the different economic backgrounds, cultural backgrounds and living conditions determined the different clothing characteristics of Ewenkis. Male and female Ewenki living in Inner Mongolia wear long robes embroidered with wide borders, however the robes worn by women have a unique charm insofar as the hem is wide and pleated. They also often wear an upturned broad-brimmed felt hat with red tassels. Both men and women wear wide leather belts and leather boots.

衬托婀娜身姿的秀美喇叭裙
Flared Skirts
　　鄂温克女子的袍装，上端紧身，下端则是如喇叭裙状，宽大，有褶，马蹄袖。不同年龄、季节及是否婚嫁，式样都有区别：未婚女子衣襟上缝有一道

或两道宽约3厘米的倒直角边形的花边，已婚妇女的袍子肩部有重叠式的起肩，并且还外套坎肩，坎肩镶彩边。

Ewenki women wear a wide and pleated gown that is tight at the top, flared at the bottom and has horseshoe sleeves. Clothing style varies according to age, season and marital status, for example, the cloth front part of garments worn by unmarried women has one or two lace strips measuring roughly three centimetres in width and at right-angles, while married women wear robes with overlapping rising shoulders and sleeveless jackets with colourful edges.

玲珑鬼马布面帽
Cloth Hats

鄂温克族的布面帽两边有长帽耳，前面有一个往上翻起的小帽耳，牧区人所戴的帽子呈圆锥形，尖端有一束红缨穗，帽面多用蓝色、天蓝色布或绸制作，帽耳的里子冬、夏季不同，冬季用羔皮或獭皮，夏季用蓝呢绒。

The hats worn by the Ewenki people have long ears on both sides and a smaller upturned one in the front. They are worn by herdsmen, are conical and there are a bunch of red tassels at the tip. The surface of the hat is made mostly of blue or sky-blue cloth or silk and the lining of the ears differs between summer and winter. Lambskin or otter skin is used in winter and blue wool is used in summer.

扫码收听
音频内容

鄂伦春族
Oroqen

　　世居中国东北的一个人口较少的民族，主要分布在内蒙古自治区东北部的鄂伦春自治旗、扎兰市和黑龙江省黑河市、塔河市、呼玛县、逊克县及嘉荫县。因住在寒冷地区善狩猎而闻名，其服饰也多与狩猎相关。男女多穿皮袍，特色服饰是一种用完整的狍子头制作的帽子，以及鱼皮靴、鱼皮绣花手套。鄂伦春人智慧地运用居住地的自然资源——狍子皮，将狍皮制品加工得多姿多彩，创造了独特的服饰文化。

　　The Oroqen people are a relatively small ethnic group living in Northeastern China. They are mostly distributed in the Elunchun Autonomous Banner and Zhalan city of Inner Mongolia and Heilongjiang province. They are known for being good at hunting in cold areas and their clothes are connected with this activity. Both men and women wear leather robes and their most characteristic garments are a hat made from a complete roe deer head, fish skin boots and embroidered fish skin gloves. The Oroqen people were able to process diverse forms and colours of roe deer products and create a unique clothing culture from their local natural resource of roe deer.

傻狍子里出珍品
Roe Deer Treasures

　　鄂伦春族传统皮衣，男女都穿，采用狍子皮制作，将狍子皮进行鞣制、染色后手工缝制而成，不仅是保暖神器，穿起来更是富贵逼人，派头十足。

　　The traditional leather clothing of the Oroqen people is worn by both men

and women and made of roe deer leather. The roe deerskin is tanned, dyed and hand-sewn. It provides warmth.

集霸气与可爱为一体的狍脑帽
Roe Deer Hat

这种帽子鄂伦春语称"灭塔哈"，用一整只狍子的头皮制成，两角耸立，男子外出狩猎时常戴，不仅保暖，还能用以伪装，是鄂伦春族最富特色的服饰之一。

The hats worn by the Oroqen people are made from a whole roe deer's scalp and the deer antlers are projected upwards. Men often wear it when they go hunting, not only because of its warmth, it also acts as a camouflage. It is one of the most distinctive costumes of the Oroqen people.

赫哲族
Hezhe

中国东北一个历史悠久的民族，主要分布在黑龙江、松花江、乌苏里江交汇的三江平原地区，即黑龙江省同江、抚远、饶河等地。世居三江原野，以捕鱼为生的赫哲族，除了宽大厚实的毛皮袍子之外，最有特色的传统服饰就是用鱼皮做的鱼皮长衫、鱼皮套裤和鱼皮靰鞡（一种皮质手工鞋，里面垫着乌拉草），因此，古代文献中关于赫哲族有"鱼皮部"等类似叫法。赫哲族妇女还喜用打磨光滑的海贝和柔软的鹿皮作装饰，镶在自染自绣的衣边上。

The Hezhe people live in Northeastern China and they have a long history. They are mainly distributed in the Sanjiang Plain, where the Heilongjiang, Songhua and Wusuli rivers meet. Living in the Sanjiang wilderness, the Hezhe people survive on fishing, and apart from large thick robes, their most distinctive traditional costumes are made of fish skin and include fish skin gowns, over-trousers and wula (which are handmade leather shoes lined with wula sedge). Because of these clothes, the Hezhe people were referred to by names similar to the *Fish Skin Tribe* in ancient literature. Hezhe women also like to use polished seashells and soft deerskin as decorations that are attached to the hems of self-dyed and embroidered garments.

鱼皮也能做衣服
Clothes Made of Fish Skin

鱼皮服，赫哲语称之为"乌提库"，是过去赫哲族的日常服装。它耐磨、抗湿、保暖，而且方便江上劳作，衍

生产品还有鱼皮套裤和鱼皮靴靰鞡。鱼皮服因为制作工艺繁复，现在的赫哲族人已很少穿着，但也是当地最热门的旅游产品。

Fish skin clothing was the daily clothing of the Hezhe people in the past. It is wear-proof, moisture-resistant, warm and convenient for working on the river. Derivative products include fish skin overtrousers and fish skin wula shoes. Hezhe people rarely wear fish skin clothing today because of the complicated crafting process involved; however, it has become the most popular tourist product in the local area.

扫码收听 音频内容

朝鲜族
Korean

　　跨境民族，中国境内主要分布在吉林、黑龙江、辽宁三省，集中居住于图们江、鸭绿江、牡丹江、松花江及辽河、浑河等流域。以轻盈明艳、翩翩宜舞为特色的朝鲜族女服，上装为笼袖短衣，下装为细褶宽裙，一短一长，一张一弛，和谐统一。胸前领条飘逸，更显灵秀。男服以肥大的"跑裤"和船型鞋为特色，上衣多为浅色，外罩一深色坎肩，显得稳重大方。

　　The Korean people living in China are mostly situated in Jilin, Heilongjiang and Liaoning provinces. Light and brightly coloured clothes are characteristic features of the clothing worn by Korean women. The clothes consist of a curved sleeve short jacket (jeogori) and a thin pleated broad skirt (chima). There is harmony and unification in both the short and long garments and tension and slackening of fabric. The beautiful collar resting over the chest drifts elegantly away. The clothing worn by Korean men is characterised by *roomy trousers* (known as baji) and boat-shaped shoes (called kkotsin). The top part (sokgui) is mostly pale in colour and covered by a dark sleeveless jacket (jeogori).

穿上"则羔利"你也是大长今
The Korean *Jeogori*

　　朝鲜族妇女传统上衣，从高丽朝开始，就流行衣长到胸下部，以飘带取代腰带，白色布条装饰领边的样式，并逐渐形成鱼肚形长袖，袖口窄，衣裙短的"则羔利"样式。

　　This traditional upper jacket

originated in the Goryeo dynasty. Originally, the jackets reached to below the chest, white strips of cloth decorated the collar and a ribbon was used to replace the need for a waist belt. All of these characteristics gradually changed to form the contemporary shorter *jeogori* which has fish belly long sleeves with narrow cuffs.

是翩翩佳郎，也是迷人欧巴
Men's Clothing

朝鲜族男子服装一直变化不大，民间传统的男装多为白色，上衣短，裤长且肥大，裤口系带，外套深色坎肩或有布袋、纽扣的斜襟长袍，足登船型鞋，鞋头翘起，头戴黑笠礼帽，礼帽以马鬃和竹条制作，到 20 世纪初变成短顶窄檐型，平民也可戴。

Korean men's clothing has not changed a lot. Traditional popular men's clothing was mostly white and consisted of short tops, long roomy trousers which were tied-shut around the ankles, a dark sleeveless jacket or a robe with a slanted buttoned lapel and boat-shaped shoes with raised fronts. A black cylindrical hat made of horse mane and bamboo strips with a broad brim was also worn. However, by the early twentieth century, the hat was lowered and brim narrowed, making it more suited to the populace.

锡伯族
Xibe

　　原居中国东北地区的古老游牧民族，清乾隆年间征调部分至新疆。现在的锡伯族多数居住在新疆察布查尔锡伯自治县和霍城、巩留等县，辽宁、吉林、黑龙江、内蒙古等省区有散居。锡伯族服饰兼有满、蒙、维等民族特色，男子头戴毡帽，身穿大襟长袍，足蹬长筒皮靴；女子着旗袍和齐腰小坎肩，围巾、腰带与蒙古族类似，头箍博采众家之长，独具特色的是缀有银泡和珠穗的王冠状头箍，十分别致美丽。

The Xibe people are an ancient nomadic people that were indigenous to the northeast region of China but some of whom were called to move to Xinjiang during the Qianlong period of the Qing dynasty. Many Xibe people now live in Qapqal Xibe autonomous county, Huocheng and Gongliu in Xinjiang; there are also some scattered settlements in Liaoning, Jilin, Heilongjiang and Inner Mongolia. Xibe costumes combine the Manchu, Mongolian and Uyghur characteristics. Men wear felt hats, long robes which are buttoned on the right, and long leather boots. Women wear qipao, tight-fitting sleeveless jackets, scarf and belts similar to those worn by the Mongol, and headbands which mix other ethnic elements. What is unique is the remarkable crown-shaped headband which is embellished with silver balls and a pearl fringe.

锡伯新娘的"王冠"
The Crown Worn by Xibe Brides

　　锡伯族青年女子结婚时头上的饰

物镶有珠带、流苏等，形似王冠。正中一般为凤凰或花卉图案的金银或铜制品，额前流苏则是成串的珍珠、琥珀或玻璃珠等，华丽高贵。

Head ornaments, which are crown-shaped and inlaid with beaded tassels, are worn by Xibe brides when they marry. The centre is usually made using gold, silver or copper, which are patterned with phoenixes or flowers and plants, and in front of the forehead is a fringe made of strings of pearls, amber or glass beads. This crown appears both gorgeous and noble.

装进香囊里的爱情
Ohojaka

锡伯族新娘的配饰奥合加卡，亦称"肢带"。功能类似于香囊，用色彩艳丽的丝织物制成，内装各种芳香物。它是锡伯族青年男女的随身携带之物，有时也作为爱情信物。

Xibe brides wear accessories called *ohojaka*, also known as limb bands. This item is made of colourful silk and its function resembles that of a fragrance bag, because it holds various kinds of scent. Young Xibe men and women carry them about and the bags are sometimes used as a token of love.

扫码收听 音频内容

壮族
The Zhuang

　　跨境民族，中国境内主要聚居在广西、广东、云南、贵州等省区。作为中国少数民族中人口最多的民族，壮族服饰也是五花八门。总体来看，壮族妇女多穿青布蜡染衣裙，有的支系较素雅，有的则喜好繁饰。云南文山壮族还有一种用蛋清等材料多次浸染加工的亮布面料做成的上衣，光洁硬挺的圆形衣襟向两侧外翘，如同飞檐，很有特色。

　　The Zhuang people living in China are mainly concentrated in Guangxi, Guangdong, Yunnan and Guizhou. As the second most populous group in China, the Zhuang people's clothing are also varied. Generally speaking, Zhuang women wear black batik dresses, with some being simpler and others more complex. The Zhuang of Wenshan in Yunnan also have a very distinctive jacket which is dyed and processed multiple times, using materials like egg whites. These jackets have smooth yet stiff rounded lapels that curling outwards.

质朴简约的男性气质
Rustic Masculinity

　　壮族男子服装各地差异不大，多为上穿对襟短衫，下着长裤，系腰带，腰带一般用长两米多的家织土布制作。有的山区年长男子还穿斜襟长衫或短衫。

　　Male Zhuang clothing does not vary much on a regional basis. It mostly consists of a short buttoned jacket, long trousers and a two-metre long belt which is made of homespun cloth. In some mountainous areas, older men wear long or short shirts with slanted lapels.

素雅头巾，浪漫的点睛之笔
Elegant Headscarves

广西壮族妇女着大襟右衽白色长袖衣，绣花围裙，黑色宽脚裤，用白色或浅色花头巾包头，交叉盘于头顶成梭形，有穗的两端自然地垂在肩上，显得清秀飘逸。

Zhuang women of Guangxi wear a long-sleeved white gown which is buttoned on the righthand side, broad black flared trousers and a white or light-coloured headscarf. The headscarf is wrapped in the shape of a shuttle and the tassels on both ends hang naturally over the shoulders.

扫码收听
音频内容

瑶族
Yao

　　跨境民族，历史非常悠久，传说是古代东方"九黎"中的一支。主要居住在广西壮族自治区、广东、云南、湖南及贵州的部分地区。不同支系的瑶族服饰有不同的特点，某些支系的名称正来源于此，如广西"白裤瑶"得名于男子齐膝白裤；"顶板瑶"源于妇女头顶的板状装饰；云南的"白头瑶"是由于妇女爱以白线缠头为饰；"蓝靛瑶"则因善染蓝靛衣而得名。关于服饰也有一些动人的传说，如云南瑶族传说，他们南迁时，曾约举芭蕉花为标识，后来演变为仿芭蕉花的尖顶红包头，称"红头瑶"；后行队伍因找不到芭蕉花，改以芭蕉杆和叶为标识，于是便演变为头饰圆柱形发箍的"平头瑶"和头顶平帕的"沙瑶"。

　　The Yao ethnic group has a very long history, according to legend, they represent one branch of the "nine li" of the ancient orient. Those in China mostly live in Guangxi Zhuang Autonomous Region, Guangdong, Yunnan, Hunan and Guizhou. The characteristic features of the Yao clothing vary according to the different branches and the names of some branches are derived from this variation in clothing, for example, *Baiku Yao* (literally: white trouser Yao) of Guangxi were named after their white shorts; *Dingban Yao* (literally: top-plate Yao) were named after the plate-shaped head ornaments worn by women; *Baitou Yao* (literally: white-head Yao) were named after the white thread women used to decorate their heads, and *Landian Yao* (literally: indigo Yao) were named as such due to their skill in dyeing clothes indigo. There are also some moving legends about costumes. One such legend is that of the Yao ethnic group of Yunnan. When they moved south, they initially choose the Chinese banana flower as their symbol, which later developed into a pointed

red turban symbolising the Chinese banana flower; this group was called *Hongtou Yao* (literally: red-head Yaos). Later on, they used Chinese banana branches and leaves as their symbol because they were no longer able to find the flowers; this meant their headwear evolved into a cylindrical headband (known as *Pingtou Yao*) and a flat hat (*Sha Yao*).

瑶族男孩的第二层皮肤
Yao Boys' White Shorts

"白裤瑶"支系的男子多穿短而紧身的白色裤子，裤管装饰五条用红丝线绣成的直条，中间一条长、两边两条短，据说这些红色直条象征其先祖为了保卫民族尊严带伤奋战的十指血痕，是纪念先祖的标志。

Male *Baiku Yao* often wear tight-fitting white shorts. The trouser legs are decorated with five straight strips embroidered with red silk thread, with the middle one being long while the others distributed evenly on either side are shorter. It is said that these red strips symbolise the ten-figure blood stains that were borne from the battles their ancestors fought to preserve dignity of the ethnic group, as such, is a mark of commemoration to their ancestors.

用来铭记历史的蜡染装饰
Batik Decorations to always Remember History

瑶族女子上衣背部的花背牌，实际上是一个正方形的蜡染图案饰物，蓝、白两色的蜡染图案上还有用橘红色丝线绣制的花纹。据说花背牌代表了当年被土官夺走的瑶王印章。

The insignia on the backs of the jackets worn by female Yao is a square batik decoration which consists of a blue and white wax patterns embroidered with reddish orange silk. It is said that the insignia represents a Yao king's seal which who was taken away by local officials.

毛南族
Maonan

　　主要分布在中国广西壮族自治区西北山区的环江、河池、南丹、都安等地，贵州也有少数聚居。毛南族服饰在历史上受到壮、汉等民族的影响，相互间有一定的吸收和补充。如毛南族妇女穿的宽长、绲边、右襟的上衣和绲边裤就和壮族有许多共同点，只是毛南族上衣有领，而壮族没有。毛南族男女服装无太大差异，均为大襟布衣布裤，唯女子喜系一围腰。特色服饰是一种名为"顶卡花"的花竹帽。

The Maonan people are mainly distributed in the northwest mountainous area of Guangxi, and some living in Guizhou. Their costumes have historically been influenced by Zhuang, Han and other ethnic groups, and they have assimilated and supplemented each other to a certain extent. For example, women's wide, long, right buttoned tops and trousers, both with lace trim edges, have many points in common with the Zhuang clothes, apart from the Maonan clothes have collars whereas the Zhuang do not. With the exception that women wear aprons, there is not much difference between the clothes worn by Maonan men and women because they all wear right buttoned tops and trousers. Their characteristic item of clothing is a floral bamboo hat, which is called *Head Flower*.

相亲必备礼物——顶卡花
The *Head Flower* – An Essential Gift for a First Date

　　毛南女子出嫁后，一般用青布包头，露出发顶，如外出劳动或走亲串寨则喜欢戴上花竹帽，此帽当地称为

"顶卡花"，一半用金竹、水竹篾手
工编制，分里外两层，帽底编花，做
工精致，是毛南族男女老少常戴的晴
雨两用帽，也是男女相恋的定情信物。

After getting married, Maonan
women usually partially wrap their heads
in green cloth with the tops of their head
left exposed. When they go out, they prefer
to wear flower bamboo hats, which are
called *Head Flower*. The hat is made of
golden bamboo and fish scale bamboo in
two layers. Flowers are added to the bottom
of the hat. The hat can be worn on both
sunny and rainy days by all people. It also
represents a token of love for couples.

扫码收听
音频内容

京族
Jing

　　跨境民族，在越南也称"越族"，中国境内主要聚居于中国广西壮族自治区南端防城港市的山心、巫头、万尾（"京族三岛"）以及潭吉等地。京族服装用料单薄，结构简洁，具有亚热带服饰的特点。京族渔女身穿无领长袖紧身衣，下着肥管长裤，也有的外套一件类似旗袍的紧身大襟长衫，喜戴尖顶斗笠，多赤脚。海风吹来，宽大的裤管和长衫衣襟随之飘动，仪态飘逸。有时，她们又将开衩的长衫两片前襟掀起，在腰部打结，形似蝴蝶。京族也常穿木屐，木屐除平时穿用外，还是缔结美满姻缘的象征物。

The Jing people in China live mostly in Shanxin, Wutou, Wanwei of Guangxi. The materials used in Jing clothing are thin, with simple structures and characteristic of subtropical clothing. Jing fisherwomen wear long-sleeved collarless tops, loose-fitting trousers, a tight-fitting long gown, which resembles the qipao as an overgarment, a pointed plaited bamboo hat and often go barefoot. The trousers and long gown flutter gracefully in the sea breeze. Sometimes they lift and knot the two slits at the front of the gown around their waist like a butterfly. They often wear clogs, an item that also symbolises a happy marriage.

穿"奥黛"的少女，海的女儿
Aodai Dress
　　京族女性上身穿窄袖、紧身、对襟或大襟的圆领短上衣，内挂一块菱形遮胸布；外出时常加穿白色、窄袖长外衣，形如旗袍，越南语称"奥黛"。

下身穿宽而长的裤子，以黑、白、红褐色为多。

Jing women wear a buttoned round-collared short top that is tight-fitting and has narrow sleeves on top of a rhombus-shaped chest cover. When they go out, they often wear white, narrow-sleeved overgarments shaped like *qipao*. These are called *aodai* in Vietnamese. In terms of bottoms, they mostly wear wide and long black, white or reddish-brown trousers.

捕鱼阿哥的专业穿戴
Fishermen's Clothes

京族男子穿袒胸的窄袖上衣，衣长及膝，腰间束带。下穿宽而长的裤子，裤管肥大。这种装束简便宽松凉爽，适合长年在海上捕鱼劳动。

Jing men wear narrow-sleeved tops that leave their chest exposed. This garment reaches knee level and is tied at the waist. Long roomy trousers are also worn. This type of dress is simple, comfortable and cool, making it suitable for fishing on the sea all year round.

扫码收听 音频内容

黎族
Li

　　海南岛的原住民，主要聚居在中国海南省，贵州也有少量分布。黎族是较早利用棉花纤维作为衣着原料的民族之一，服饰色彩主要以黑色或蓝色为底，其上配以各种颜色的镶饰及纹样。黎族服饰纹样常有反映狩猎、婚嫁和祭祀的内容，以及人们载歌载舞、欢庆丰收的情景，具有浓郁的民俗色彩。黎族女子一般上着窄袖紧身短衣，前襟敞开，露出素净的胸衣，颈上戴多重银项圈，有的还在双耳饰以两串大耳圈，带有一种原始野性的美感。男子服装以包头、短衣和短裤为主。

The Li people are the indigenous people of Hainan island and they mostly inhabit this island with some distributed in Guizhou. The Li were one of the earliest ethnic groups to use cotton fibre as a raw material for clothing and blue or black is used as the main base colour which is decorated with various colours of inlays and patterns. The patterns on the Li clothing have strong folklore connotations insofar as they often make known the contents of hunting, marriage and sacrifice, as well as the scenes of festive singing and dancing, and celebrating the bumper harvests. The Li women generally wear open short narrow-sleeved tight-fitting tops over a plain corset. Several kinds of silver neckbands are worn around the neck and some women even wear a pair of large earrings. Men mostly wear turbans, short tops and shorts.

男生也穿"迷你裙"
Even Men Wear Mini-skirts
黎族男子多结髻缠头，上衣为无

领对襟短衫，下穿由前后两幅布组成的"吊檐"（裹裙的一种）或裤子，凉快又清爽，正适合海南炎热的天气。

Many Li men tie their hair in a bun, wear a short collarless buttoned shirt and either a kind of wrapping skirt, which is made of two strips of fabric worn in front and behind, or trousers. These clothes help to keep the wearer cool in the scorching heat of Hainan.

别具一格的热带野性气质
A Unique and Tropical Style

黎族妇女一般束髻于脑后，头披绣花头巾，上衣多为对襟无纽式样，衣尚黑色，下穿有中长短之分的无褶织绣筒裙。盛装时戴项圈、手镯、脚环、耳环等，隆重且典丽。

The Li women usually wear their hair in a bun behind their heads and wear embroidered headscarfs on top. Their tops mostly consist of buttonless blouses and black is preferred. A non-pleated woven and embroidered skirt is divided into long, short and medium lengths. During special occasions, neckbands, bracelets, anklets and earrings are worn, giving their attire a ceremonious and grand quality.

扫码收听 音频内容

畲族
She

中国南方游耕民族,分布于福建、浙江、江西、广东、安徽等省的部分山区,以福建福安和浙江云和最为集中,与汉族交错杂居。畲族妇女上着黑色大襟衣,衣领处绣马牙状花纹,下穿黑色长裤,足蹬方头黑布厚底鞋,腰系围裙,围裙上端两角绣有花纹。男装为短衣长裤,对襟坎肩,爱用艳色装饰襟边。

The She, a nomadic people of southern China, are mostly distributed across some mountainous areas of Fujian, Zhejiang, Jiangxi, Guangdong, Anhui. They are mostly concentrated in Fu'an of Fujian and Yunhe of Zhejiang, two places where they live alongside the Han people. The She women wear large black blouses, which have collars embroidered with patterns in the shape of horse teeth, black trousers, square-toed black cloth platform shoes and aprons that have embroidered patterns in the top corners. Men wear short tops and trousers, sleeveless jackets which have bright colours decorating the hems.

穿上"凤凰装",做自己的女王
The Phoenix Costume

畲族妇女服饰以"凤凰装"最具特色。相传美丽的凤凰带来了五彩斑斓的凤凰装,红绳盘头象征着凤冠;全身佩挂叮当作响的银首饰,象征凤凰的鸣啭。

She women's clothing is characterised by the *phoenix costume*. According to legend, a beautiful phoenix brought a brightly-coloured costume and a red rope headdress symbolising the

phoenix's crown. Silver adornments, which cover the body, jingle to symbolise the chirping of the phoenix.

高山族
The Gaoshan

主要聚居在台湾中部山区、东部纵谷平原和兰屿岛上。高山族在台湾本土有多种支系和称谓，服饰也因之差异较大。由于气候较热，他们的服装一般较短，传统服装有宽肩无袖对襟或侧斜襟上衣，椰皮背心，对襟短袖长衫和筒裙等样式，色彩明艳和谐，尤其喜欢用贝壳、兽骨、羽毛、珍珠等装饰头部和身上，颇有原生态风格。特别是排湾人独有的琉璃珠，系贵族祖传宝物，是佩戴者身份、地位和财富的象征，在婚礼或祭祀等场合，尤其显出其价值。

The Gaoshan people mostly inhabit the mountainous regions in the centre, the eastern valleys and plains, and Lanyu island in Taiwan. The Gaoshan ethnic group have many branches and appellations, because of this, the costumes vary greatly. Their clothes are usually quite short because of the hot weather. Traditional clothing includes several harmonious yet brightly coloured styles including wide-shouldered sleeveless shirts or side-slanting tops, coconut skin vests, long gowns with short sleeves and skirts. They especially like to decorate the head and body with shells, animal bones, feathers and pearls, giving them a primordial and ecological style. In particular, the glass beads that are unique to the Paiwan people are ancestral treasures passed down by the nobility; they are a symbol of the wearer's identity, statues and wealth, the value of which is revealed at weddings or sacrifices.

快乐奔放的阳刚之美
Masculinity Decoration

高山族阿美男子上身常穿颜色鲜艳的无袖褂子，胸部和背部绣有几何形的花纹。前开襟，胸前斜跨一根带

有装饰物的条带，喜欢戴耳饰，节庆时也戴插有羽毛兽骨的头饰。

Gaoshan Amei men often wear brightly-coloured sleeveless gowns which have geometric patterns embroidered on the chest and back. The front of their gowns is worn open and a strip with ornaments is worn diagonally across the chest. They like to wear earrings and they wear headdresses which are decorated with feathers and animal bones during festivals.

华丽男帽——地位的标志
The Magnificent Hat

高山族阿美人男帽以细竹制成帽子的骨干，再以纸板固定于骨架上，上贴缝红色布片，并缀有各色的绒毛球、亮片与塑胶珠，帽上端插羽毛做装饰，下缀流苏。

The backbone of the hat worn by male Amei(a branch of Gaoshan ethnic group) is made of fine bamboo and cardboard is fixed to the skeleton. A piece of red cloth is attached which is decorated with various colourful fur balls, sequins and plastic beads. The top of the hat is decorated with feathers and the bottom with tassels.

朱色流转，是热情似火的霓裳
Colourful Tops for Women

阿美妇女平时上身穿对襟长袖短衣，配修身短裙，颜色以鲜艳为主。胸前斜挂一块绣有图案的方形布，节庆时佩戴用玛瑙及珍珠串成的头饰和项链。

Amei women usually wear brightly-coloured clothes consisting of a long-sleeved short shirt together with a slim-fitting short skirt. A square cloth embroidered with patterns is worn diagonally across the bosom and, during festivals, headdresses and necklaces made of agate and pearls are worn.

有装饰物的条带，喜欢戴耳饰，节庆时也戴插有羽毛兽骨的头饰。

Gaoshan Amei men often wear brightly-coloured sleeveless gowns which have geometric patterns embroidered on the chest and back. The front of their gowns is worn open and a strip with ornaments is worn diagonally across the chest. They like to wear earrings and they wear headdresses which are decorated with feathers and animal bones during festivals.

华丽男帽——地位的标志
The Magnificent Hat

高山族阿美人男帽以细竹制成帽子的骨干，再以纸板固定于骨架上，上贴缝红色布片，并缀有各色的绒毛球、亮片与塑胶珠，帽上端插羽毛做装饰，下缀流苏。

The backbone of the hat worn by male Amei(a branch of Gaoshan ethnic group) is made of fine bamboo and cardboard is fixed to the skeleton. A piece of red cloth is attached which is decorated with various colourful fur balls, sequins and plastic beads. The top of the hat is decorated with feathers and the bottom with tassels.

朱色流转，是热情似火的霓裳
Colourful Tops for Women

阿美妇女平时上身穿对襟长袖短衣，配修身短裙，颜色以鲜艳为主。胸前斜挂一块绣有图案的方形布，节庆时佩戴用玛瑙及珍珠串成的头饰和项链。

Amei women usually wear brightly-coloured clothes consisting of a long-sleeved short shirt together with a slim-fitting short skirt. A square cloth embroidered with patterns is worn diagonally across the bosom and, during festivals, headdresses and necklaces made of agate and pearls are worn.

参考文献

[1]   宋兆麟,高可,张建新.中国民族民俗文物词典 [M].太原:山西人民出版社,2004.9.

[2]   钟茂兰,范朴.中国少数民族服饰 [M].北京:中国纺织出版社,2006.8.

[3]   杜钰洲,缪良云.中国衣经 [M].上海:上海文化出版社,2000.4.

[4]   徐海荣.中国服饰大典 [M].北京:华夏出版社,2000.1.

[5]   首届中国民族服装服饰博览会.中国民族服饰博览 [M].昆明:云南人民出版社,2001.9.

[6]   杨振生.茶马古道上的民族服饰 [M].昆明:云南人民出版社,2018.11.

[7]   云南新闻图片社,云南省群众艺术馆,云南省歌舞团.东方彩霞——中国五十六个民族服饰 [M].昆明:云南民族出版社,1996.12.

[8]   云南省群众艺术馆.云南民族民间艺术 [M].昆明:云南人民出版社,1994.2.

[9]   李昆声,周文林.云南少数民族服饰 [M].昆明:云南美术出版社,2002.1.

[10]   中国西部少数民族服饰 [M].成都:四川教育出版社,香港国际出版社,1993.3.

[11]   李肖冰.中国西域民族服饰研究 [M].乌鲁木齐:新疆人民出版社,1995.8.

[12]   王江红.云南民族传统服饰 [M].昆明:云南人民出版社,2016.12.

[13]   韦荣慧.中华民族服饰文化 [M].北京:纺织工业出版社,1992.9.

[14]   周少华.中国白裤瑶民族服饰 [M].北京:化学工业出版社,2017.6.

[15]   贵州省文化厅.苗系列画册 [M].北京:人民美术出版社,1992.7.

[16]   贵州省科技教育领导小组办公室,贵州省民族事务委员会.贵州世居少数民族服饰经典 [M].贵阳:贵州民族出版社,2013.7.

[17]   张柏如.侗族服饰艺术探秘(上)服饰篇,汉声杂志70[J].台北:汉声杂志社,1994.10.

[18]   方钧玮.原住民织品及饰品图录 [M].台北:国立台湾史前文化博物馆,2001.12.

[19]   陈高华,徐吉军.中国服饰通史 [M].宁波:宁波出版社,2002.10.

[20]   Dai Ping,Ding Jiasheng.ETHNIC COSTUMES AND CLOTHING DECORATIONS FROM CHINA[M].Chengdu:Sichuan People's Publishing House,1989.